ADVANCES IN METAHEURISTICS

Applications in Engineering Systems

ADVANCES IN METAHEURISTICS

Applications in Engineering Systems

Timothy Ganesan, PhD
Pandian Vasant, PhD
Irraivan Elamvazuthi, PhD

CRC Press
Taylor & Francis Group
Boca Raton London New York

CRC Press is an imprint of the
Taylor & Francis Group, an **informa** business

CRC Press
Taylor & Francis Group
6000 Broken Sound Parkway NW, Suite 300
Boca Raton, FL 33487-2742

Printed on acid-free paper
Version Date: 20161021

International Standard Book Number-13: 978-1-4987-1548-5 (Paperback)

Library of Congress Cataloging-in-Publication Data

Names: Ganesan, Timothy, author. | Vasant, Pandian, author. | Elamvazuthi, Irraivan, author.
Title: Advances in metaheuristics : applications in engineering systems / Timothy Ganesan, Pandian Vasant, Irraivan Elamvazuthi.
Description: Boca Raton, FL : CRC Press, Taylor & Francis Group, LLC, CRC Press is an imprint of Taylor & Francis Group, an Informa Business, [2017] | Includes bibliographical references.
Identifiers: LCCN 2016030313| ISBN 9781498715485 (acid-free paper : hardcover) | ISBN 9781498715492 (electronic)
Subjects: LCSH: Industrial engineering--Mathematics. | Electric power systems--Mathematics. | Materials science--Mathematics. | Mathematical optimization. | Heuristic algorithms.
Classification: LCC T57 .G36 2017 | DDC 620.00285/63--dc23
LC record available at https://lccn.loc.gov/2016030313

Visit the Taylor & Francis Web site at
http://www.taylorandfrancis.com

and the CRC Press Web site at
http://www.crcpress.com

Contents

Foreword

The main purpose of this book *Advances in Metaheuristics: Applications in Engineering Systems* has been to provide details on current approaches utilized in engineering optimization. This work provides a comprehensive background on metaheuristic applications, focusing on core engineering fields related to, for example, "energy," "process," and "materials." With the help of this book, readers can learn to solve real-world engineering optimization problems effectively through appropriate techniques from emerging fields including mathematical optimization, evolutionary and swarm intelligence, and multiobjective optimization. The book also inspires for future research and applications.

The book is organized in a modular way based on chapters which are grouped into three parts. The three authors altogether introduce advanced results in the areas of *metaheuristics*, their basic elements, methods, applications, and directions. In fact, these have become very important methodologies in practical utilizations from the perspectives of an effective energy supply to consumers, of control theory on electrical and electronical devices, and of modern materials. An effective operation of industries, of the economy, and of the life of the people strongly depends on these areas. This book herewith, through its collection and exposition of new state-of-the-art trends, has turned out to be a truly worthwhile project. It offers a precious and useful service for the academic sector, for the real-world practice of engineering and economics and, finally, for developing policies, strategies, and solutions to enhance the living conditions and perspectives of the people on earth.

In recent decades, the methodologies, toolboxes and "kit-boxes" of engineering, economics, computer science, applied mathematics, statistics, informatics, and bio- and life-sciences, have gained the interest of a vast number of researchers and practitioners from all over the world, in emerging analytics, algorithms, and information technologies, giving a strong impact to all areas of traditional and information technologies, and also in economics, finance, and social sciences. Hereby, a central role is played by optimization, optimal control, and probability theory, in theory, methods, and applications. The presented book gains from that fast growth. We would like to mention that, in future investigation, even optimal control of stochastic hybrid systems could be further addressed, where those systems allow for an inclusion of jumps or impulses, regime switches, and paradigm shifts.

This valuable compendium addresses the following purposes and goals, research topics, and techniques:

- It provides details on the implementation of metaheuristic techniques in engineering problems.
- It explores multiobjective optimization scenarios and their solution strategies in engineering applications.
- It supplies the reader with details on the characteristics and behaviors of metaheuristic algorithms when applied to engineering.

- It presents a guide for the engineer/decision maker for selecting effective metaheuristic techniques when faced with specific engineering problems.
- It offers concepts regarding improvement methods for enhancing the performance of metaheuristics.

The presented book aims at three clusters of subjects, namely:

Part I: Energy Systems, including

- *Geometric Optimization of Thermoelectric Coolers*,
- *Mean-Variance Mapping Optimization for Economic Dispatch*, and
- *Smart Charging Optimization of Plug-in Hybrid Electric Vehicles*

Part II: Process Optimization, including

- *Three-Phase Separator Optimization Using Bacteria Foraging*,
- *Plant Optimization Using Artificial Fish Swarm Algorithm*, and
- *Multiobjective Optimization of Bioethanol Pretreatment Process*

Part III: Material Engineering, including

- *Bioactive Compound Extraction Process Optimization*,
- *Multiobjective Optimization of Cement-Bonded Sand Mold System*,
- *Multiobjective Optimization of Green Sand Mold Product*, and
- *Multiobjective Optimization of Resin-Bonded Sand Core Properties*.

This indicates that a rich and varied discussion is unfolded along the chapters.

The authors of this book's chapters are experienced and enthusiastic researchers and scholars from all over the globe, who associate, refine, and apply both with less model-based but more model-free and data-driven approaches of engineering and computer science and with the deep model-based techniques and methodologies of mathematics. The first approaches are sometimes called as smart or intelligent algorithms; they have roots in technological disciplines, computer science, informatics, and bio- and nature-inspired approaches of thinking and reasoning. Highly demanding and amazing challenges do exist in all fields of our modern lives, in high-tech and economies, in the areas of development, sustainability, and of gaining and implementing future chance and perspectives for humankind. In this course, there is no doubt that the areas investigated and displayed in this book, especially on energy, processes, and materials, and also the related areas of climate change, environmental protection, and guarantee and maintenance of fairness and justice, security, and liberty, play a vast role which cannot be overestimated.

Special attention paid to in this work and, as we trust, in following future work is the presence of *uncertainties* of various kinds, the high complexity of analyzing and interpreting large data sets, and entire problems that we nowadays find nearly everywhere. In this present book and in future studies inspired by it, the above-mentioned two main academic approaches, cultures, and traditions, namely, the more model-free and data-driven one from engineering and the more model-based one from mathematics, cannot be considered as separate or disjoint from each other, but from their joint promise of common chances, of synergies and of a potential use to humankind.

To all authors of the chapter contributions, we extend our appreciation and our gratefulness for their devotion, expertise, and insights which they shared with the whole academic family and with humankind. We are thankful to the publisher CRC Press and to the authors, Timothy Ganesan, Pandian Vasant, and Irraivan Elamvazuthi, for having prepared and maintained the chance and platform for experts to publish their fundamental and emerging advances and proposals. To them all we convey our sincere thanks for having made possible a premium book of a high intellectual level and, as we hope, of an academic, real-world, and human impact.

We wish you and each one of us a great pleasure and gain when reading this new work by CRC Press/Taylor & Francis, a great gain by it in all personal, professional, and social respects.

Gerhard-Wilhelm Weber*
Institute of Applied Mathematics, METU, Ankara, Turkey

Emel Savku
Institute of Applied Mathematics, METU, Ankara, Turkey

Erik Kropat
University of the Bundeswehr Munich, Neubiberg, Germany

* Further affiliations: Faculty of Economics, Management and Law, University of Siegen, Germany; School of Science, Information Technology and Engineering, Federation University Australia; Center for Research on Optimization and Control, University of Aveiro, Portugal; Universitas Sumatera Utara, Medan, Indonesia.

Preface

Since the initial stages of industry, engineers have always encountered challenges related to the enhancement of system efficiency, design, and reliability. For finding solutions to such optimization problems, engineers often employ tools from the mathematical disciplines such as convex, integer, and quadratic programming. With the recent advent of increased computational power, these techniques could be easily and readily executed when required. Besides, various artificial intelligence (AI)-based computational techniques have been developed and utilized for solving engineering optimization problems. These techniques have been proven to be effective and algorithmically efficient in solving such problems. AI-based techniques are approaches that have higher learning and self-improvement capabilities. These capabilities are employed for searching optimal solutions. These techniques encompass a wide range of metaheuristic approaches such as swarm intelligence and evolutionary algorithms. In addition, other AI-based techniques such as machine learning algorithms are also gradually being used for solving optimization problems.

Optimization problems in engineering often contain high levels of various types of complexities. For instance, they are often found to be nonlinear, nonconvex, multiobjective, and consisting of multiple nonlinear constraints. Although heuristic (or experiential) approaches are commonly employed in industrial settings, the effectiveness of these approaches may be improved by coupling them with computational techniques. This is because human intelligence per se is not efficient in performing tedious and exhaustive tasks such as rigorously searching for optimal solutions to a problem that contains various complexities. Thus, such problems are more effectively tackled if human heuristics and computational approaches go hand in hand.

The central idea of this book involves the implementation details of metaheuristic strategies to real-world engineering systems. Therefore, this book emphasizes the engineering applications and discusses the details of the metaheuristic implementation in each case. The primary motivation to write this book springs from the current lack of literature which familiarizes readers with details on the implementation of metaheuristics in engineering applications. Besides, very few literature studies focus on the information and characteristics of the optimization strategies when applied to engineering systems. In this spirit, this book provides a comprehensive background on metaheuristic applications focusing on three key engineering areas: energy systems, materials engineering, and process optimization. Besides, insights on the implementation of metaheuristic strategies to multiobjective optimization problems and information regarding methods for solution evaluation in such cases are provided. Topics such as algorithmic enhancements and performance measurement approaches (for gauging the optimized solution quality) are also covered. With this book, readers would thus be able to solve real-world engineering optimization problems effectively using the appropriate metaheuristic techniques. Hence, this book is aimed to be a guide to engineers/decision makers for selecting and implementing metaheuristics to specific engineering problems.

Part I discusses three industrial applications in the energy sector. The first chapter consists of the design optimization of thermoelectric coolers (TECs) while the second chapter involves the optimization of economic load dispatch, and the third chapter is on plug-in-hybrid electrical vehicles. In this part, details on the implementation of metaheuristics such as differential evolution (DE) and simulated annealing (SA) are provided.

Chapters 4 through 6 fall under Part II of this book: process optimization. This part contains three engineering applications which are: optimization of a three-phase separator, process plant, and a pretreatment process. In this part, various evolutionary and swarm intelligence-based strategies are analyzed. Besides, ideas involving the benchmarking of multiobjective solutions as well as approaches for measuring solution characteristics are given.

Part III contains four industrial applications in materials engineering. Here, the real-world applications are mainly focused on sand mold systems; the key concept presented in these applications is the optimization of sand mold casting processes and resin-bonded sand core properties. In these chapters, discussions involving the improvement of algorithmic characteristics via strategic algorithmic enhancements are provided.

Insights on strategies in industrial optimization have been provided by many prominent books, technical articles, and Internet resources. These resources appear in the reference section of each chapter. The interested reader may refer to these resources for further details regarding the optimization strategies, model formulations, or numerical experiments. Alternatively, many of these details are also provided by various web resources.

Acknowledgments

The authors would like to thank the Faculty of Engineering and the Faculty of Science and Information Technology, Universiti Teknologi PETRONAS, Malaysia, for their support throughout the development of this book. Timothy Ganesan would like to thank the School of Chemical Engineering at the University of Adelaide, Australia, for extending their resources to aid in the writing of this book. This book would not have made it to its final form without the help of our close friends, colleagues, students, and most importantly our families. The authors are very grateful to Khoa Truong Hoang, Imran Rahman, and Đoàn Vị Kim Khanh from the Department of Fundamental and Applied Sciences and Ho Joon Heng and Wafa Ali Soomro from the Department of Electrical and Electronic Engineering at Universiti Teknologi PETRONAS, Malaysia for their excellent research efforts and critical insights on some of the material presented in this book. Timothy Ganesan would like to thank Ms. Sima Alian and Mr. Suren Lim Sinnadurai for their patience and continual support during the writing of this book. Special thanks from Tim to Mr. Nazmi Zaidi and Mr. Noraz Khairi for helping him think through the concepts related to the development of this book. Tim is particularly grateful to his family, specifically his grandmother and aunts, for being there for him. Tim apologizes to his grandmother and aunts for insisting that they pick up some engineering to better understand his work in general and with regard to this book. Special thanks go out to our editors and personnel at CRC Press/Taylor & Francis: Jennifer Ahringer, Cindy Carelli, Michele Smith, Thomas Storr, and Kyra Lindholm.

Authors

Timothy Ganesan is a research engineer at Tenaga Nasional Berhad (TNB—Research), Malaysia. Holding a doctorate in process optimization (chemical engineering), his work focuses on power generation, systems engineering, and optimization.

Pandian Vasant is a senior lecturer at the Department of Fundamental and Applied Sciences, Faculty of Science and Information Technology, Universiti Teknologi PETRONAS in Malaysia. He holds a PhD in computational intelligence, an MSc in engineering mathematics, and a BSc (Hons) in mathematics. His research interests include soft computing, hybrid optimization, holistic optimization, innovative computing and applications. He has authored or coauthored research papers and articles in national and international journals and conference proceedings and papers. He has served as lead guest editor for three special issues of journals, edited books and conference proceedings, written book chapters, and conference abstracts, and he gave the keynote lecture for the 2nd EAI International Conference on Nature of Computation and Communication, March 17–18, 2016, Rach Gia, Vietnam. The journal, *Applied Soft Computing* (Elsevier), recognized Dr. Vasant as its top reviewer in 2009 and as an outstanding reviewer in 2015. He has 25 years of working experience at various universities from 1989 to 2016. Currently he is editor-in-chief of *International Journal of Computing & Optimization, Industrial Engineering & Management, International Journal of Swarm Intelligence and Evolutionary Computation, International Journal of Energy Optimization & Engineering*, and managing editor of the *Global Journal of Technology and Optimization.*

Irraivan Elamvazuthi obtained his PhD from the Department of Automatic Control and Systems Engineering, University of Sheffield, UK, in 2002. He is currently an associate professor at the Department of Electrical and Electronic Engineering, Universiti Teknologi PETRONAS, Malaysia. His research interests include control, robotics, mechatronics, power systems, and biomedical applications.

Part I

Energy Systems

1 Geometric Optimization of Thermoelectric Coolers

Thermoelectric coolers (TECs) are applied in a wide range of thermal energy systems. This is due to their superior features where no refrigerant and dynamic parts are needed. TECs generate no electrical or acoustical noise and are environmentally friendly. Over the past decades, many research studies were conducted to improve the efficiency of TECs by enhancing the material parameters and design parameters. The material parameters are the most significant, but they are restricted by currently available materials and module fabricating technologies. Therefore, the main objective of TEC design is to determine a set of design parameters for the purpose of enhancing its performance. Two elements that play an important role when considering the suitability of TEC in applications are: cooling rate and coefficient of performance (COP). In this chapter, the technical issues of single-stage TECs (STECs) and two-stage TECs (TTECs) are discussed. Consequently, two methods for optimizing the dimensions of the TEC (simulated annealing [SA] and differential evolution [DE]) are introduced and applied to search for the optimal design parameters of both TEC types.

1.1 INTRODUCTION

Measurement while drilling (MWD) is a system developed to perform drilling-related measurements downhole and transmit information to the surface while drilling a well (Scherbatskoy, 1982). A typical structure of the high-temperature MWD tool is shown in Figure 1.1.

An MWD system can measure several variables such as natural gamma ray, directional survey, tool face, borehole pressure, temperature, vibration, shock, and torque. Maintaining optimal payload temperatures in a typical downhole environment of 230°C requires the MWD cooling system to be capable of pumping a significant load. This requires a low thermal resistance path on the heat rejection side (or hot side). The application in an extreme environment of high temperature, high pressure, mechanical shock, and vibration requires the use of high temperature TEC materials and assemblies. Cooling of electronic components inside the MWD housing is crucial for maintaining optimal operating conditions. It has been identified that this can be accomplished using thin-film TEC devices.

TEC is a solid-state cooling device that uses the Peltier effect through p-type and n-type semiconductor elements (unlike vapor-cycle-based refrigerators) (Goldsmid, 2009). TEC is used to convert electrical energy into a temperature gradient. TEC uses

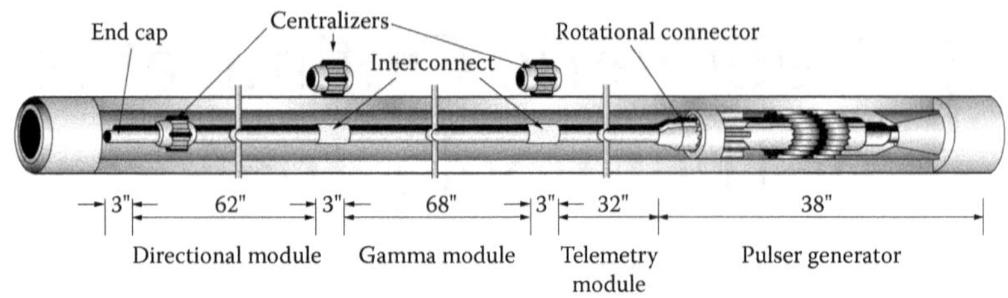

FIGURE 1.1 High-temperature MWD tool (Scherbatskoy, 1982).

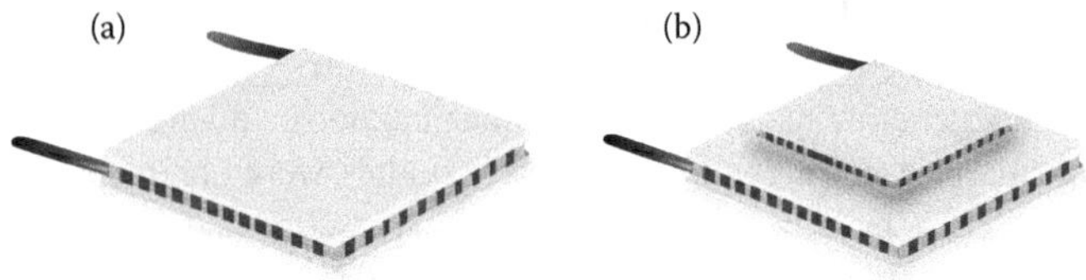

FIGURE 1.2 Single-stage and two-stage thermoelectric cooler (Kim & De Weck, 2006).

no refrigerant and has no dynamic parts, which make these devices highly reliable and require low maintenance. TECs generate no electrical or acoustical noise and are ecologically clean. TECs are compact in terms of size, are lightweight, and have high precision in temperature control. However, for this application, the most attractive feature of the TECs is that they have the capacity for cooling instruments such as MWD under extreme physical conditions. TECs can be single stage or multistages. Figure 1.2 shows STEC and TTEC.

The commercially available STECs, which can be seen in Figure 1.2, can produce a maximum temperature difference of about 60–70 K when the hot side remains at room temperature (Goldsmid, 2009). Nevertheless, when a large temperature difference is required in some special applications, the STEC would not be qualified for implementation. To enlarge the maximum temperature difference of TECs, another type of TEC which is a TTEC (Figure 1.2b) or a multistage TEC can be used (Karimi, Culham, & Kazerouni, 2011). Thermoelectric modules generally work with two heat sinks attached to their hot and cold sides in order to enhance heat transfer and system performance. The application of TECs has been partitioned by their relatively low energy conversion efficiency and ability to dissipate only a limited amount of heat flux. The two parameters that play a crucial role in characterization of TECs are the maximum cooling rate and the maximum COP. TECs operate at about 5–10% of Carnot cycle COP, whereas compressor-based refrigerators normally operate at more than 30%.

Metaheuristics are intelligent techniques that can be used for engineering design optimizations (Deb, 2001). Until now, Genetic Algorithm (GA) (Davis, 1991) and the extended version of GA, which is Non-dominated Sorting Genetic Algorithm (NSGA-II) (Deb, Pratap, Agarwal, & Meyarivan, 2002; Zhang & Hansen, 2009), are two metaheuristic optimization techniques which have been applied for optimizing the performance of both types of TECs. However, their behavior and search capability have not been analyzed deeply.

The need for TECs to be utilized as a cooling mechanism for instruments in extreme environments especially in thermal energy and gas drilling operations could be overemphasized. However, setbacks such as the relatively low energy conversion efficiency and the ability to dissipate only a limited amount of heat flux may seriously injure the performance and efficiency of these devices.

The performance evaluation of the applied metaheuristic such as GA and NSGA-II in TECs are limited. Hence, such issues need to be investigated and analyzed in more detail based on some defined criteria such as measuring the stability, reliability, robustness, and so on.

Similar sophisticated techniques in metaheuristics such as SA (Van Laarhoven & Aarts, 1987; Rutenbar, 1989; Wah & Chen, 2000) and other evolutionary algorithms like GA (Davis, 1991) and DE (Storn & Price, 1995; Mezura-Montes, Velázquez-Reyes, & Coello Coello, 2006; Ali, Pant, & Abraham, 2009) are used in their pure and hybridized form to enhance the effectiveness of the TEC model. Besides taking into account the large number of variables, these techniques optimize the model while considering the complex physical and geometrical properties of the TEC.

Although DE is in the same family of algorithms as GA, it has been claimed to have better performance as compared to GA in many applications. SA has the capability to escape from the local optimas to obtain global optimality. Therefore, in this chapter, metaheuristic techniques SA and DE are applied together with the STEC and TTEC models on a MATLAB® platform. In that system, the objectives are the cooling rate and COP. They are optimized individually or simultaneously using single-objective optimization (SOO) or multiobjective optimization (MOO) methods, respectively. The design variables, their constraints, and other fixed parameters need to be identified. The performance of each technique is strategically evalu ated based on the simulation results.

1.2 OVERVIEW OF TEC OPTIMIZATION

The main drawbacks of TECs are poor COP value and cooling rate. These issues could be improved individually or simultaneously. The parameters in the formulation of cooling rate and COP can be classified into three categories: specifications, material properties, and design parameters (Rowe, 1995). The specifications are: the operating temperatures (T_c and T_h), the required output voltage (V), current (I), and power output (P). The specifications are usually provided by customers depending on the requirement of a particular application.

The material parameters are restricted by current materials and module fabricating technologies. Due to the effectiveness of material properties on the performance of TECs, many research efforts were directed during the past 10 years toward finding a new material/structure for usage in highly efficient cooling applications and energy conversion systems. Bismuth telluride (Bi_2Te_3) is one of the best thermoelectric materials with the highest value figure of merit (Yamashita & Sugihara, 2005). Much effort has been made to raise the ZT value (which is the measure of a material's capability to efficiently generate thermoelectric power) of bulk materials based on Bi_2Te_3 by doping or alloying other elements in various fabricating processes. However, in these efforts, it was found that the *ZT* value was not more than 1. Thus, the *ZT* value was not sufficient to improve the cooling efficiency significantly. The reason is the difficulty in increasing the electrical conductivity or Seebeck coefficient without increasing the thermal conductivity (Rodgers, 2008). Recent advancements in improving *ZT* values include the work of Poudel et al. (2008), who achieved a peak *ZT* of 1.4 at 100°C from a bismuth antimony telluride (BiSbTe) p-type nanocrystalline bulk alloy. This material is an alloy of Bi_2Te_3 and is made by hot pressing nanopowders which are ball-milled from crystalline ingots. The *ZT* value of this alloy is about 1.2 at room temperature and peaks at about 1.4 at 100°C, which makes these materials very useful for microprocessor cooling applications. Consequently, the main objective of the TEC design is to determine a set of design parameters which meet the required specifications or create the best performance at a minimum cost. Table 1.1 shows the optimization techniques which have been applied in the optimization of TECs.

TABLE 1.1
Optimization Techniques Implemented for TEC Improvement

Author	Application	Advantages and Disadvantages
Cheng and Lin (2005)	GA is applied to optimize geometric design of STEC	GA proves fast convergence speed and effective search but parameter setting of GA is not suggested or discussed.
Cheng and Shih (2006)	GA is applied to optimize design parameter of TTEC	GA was applied successfully to solve a heavier problem of TTEC but the robustness of GA when applying for TTEC model is not very deterministic.
Huang, Wang, Cheng and Lin (2013)	Apply conjugate-gradient method to optimize geometric design of STEC	Effects of applied current and temperature on the optimal geometry are discussed. Performance analysis of the technique applied for STEC was not conducted. The base area of STEC is small and not practical.
Nain, Sharma, and Giri (2010b)	Using NSGA-II to optimize geometric design of STEC	MOO is performed by optimizing cooling rate and COP simultaneously. Parameters of NSGA-II were chosen based on the author's experience but the obtained results are not reliable because of the unstable performance of the algorithm.
Venkata Rao and Patel (2013)	Using teaching–learning-based optimization (TLBO) to optimize design parameters of TTEC	MOO is performed. The performance of TLBO is evaluated and compared to GA, PSO by using some defined criteria. However, parameter selection for TLBO was not implemented.

Using SOO, Cheng and Lin (2005) were the first authors implementing metaheuristics (GA) to the STEC model for finding optimal geometric designs. The geometric properties of semiconductor elements are: the total number of units, the length of the unit, and the unit cross-sectional area. These variables were optimized simultaneously to reach the maximum cooling rate under few constraint conditions. Three case studies were performed under various applied currents (I) and various temperature differences (ΔT) to evaluate the effectiveness of the search for generating better design data as compared to the available design on the market. The GA algorithm was described and applied in the TEC model. However, there were many parameters in GA (such as total number of generations, crossover, or mutation coefficient) that needed to be defined. In another study (Cheng & Shih, 2006), these authors also used GA for maximizing separately the cooling rate and the COP but for the TTEC model. In this research, two different types of TTEC which had different types of applied current were considered. Unlike the geometric design of STEC, different types of design parameters of TTEC were used as variables. They supplied current to the hot stage and cold stage of all the TTEC units. By considering the effect of thermal resistance, the mathematical modeling of TTEC was more complicated than that of STEC (Xuan, Ng, Yap, & Chua, 2002). GA was then applied successfully to solve the defined problems of TTEC. Similar to Cheng and Shih (2006), the parameter selection of GA was not described. The authors claimed that GA had a robust behavior and effective search ability. However, the robustness of an algorithm is based on the optimal capability of handling diverse problems via the changing of parameter settings. Therefore, optimizing the design variables with respect to the maximum cooling rate (and comparing against analytical results) is not sufficient to obtain a clear picture on the robustness of the algorithm.

Huang et al. (2013) developed an optimization approach which integrates a complete multiphysics STEC model and a simplified conjugate-gradient method (nonmetaheuristic). Similar to Cheng and Shih (2006), the geometric properties of the STEC were optimized to reach the maximum cooling rate under a wide range of operating conditions (temperature differences and applied current variations). The effects of applied current variations and temperature differences on the optimal geometry are discussed. The author came up with a new better design than the initial one by using the conjugate-gradient method. However, the performance analysis of the conjugate-gradient technique applied to the STEC model was not complied. And the combined optimization was proven effective only for the miniature TEC with the base area of 9 mm^2, which is smaller than the design of Cheng and Lin (2005) with 100 mm^2.

By using MOO, Nain et al. (2010a) used another version of GA, which is the NSGA-II for solving the STEC design problem. The values of the geometric properties of the STEC were optimized to achieve Pareto-optimal solutions of cooling rate and COP values at different thermal resistances. The benefits of obtaining the Pareto-frontier is that the balance between cooling rate and COP may be achieved. This benefits the designer during the selection of suitable STEC design in practical applications. However, the obtained results are not reliable since there exists a significant difference in the range of optimal design values between the results obtained using SOO and MOO. This is caused by the instability of the algorithm. In that work, the parameters of NSGA-II algorithm were chosen heuristically.

Venkata Rao and Patel (2013) used teaching–learning-based optimization (TLBO) for optimizing their design parameters of two types of TTEC. TLBO is based on the effects of the influence of a teacher on the output of learners in a classroom setting. In Venkata Rao and Patel (2013), TLBO was modified to increase the exploration and exploitation capacity. The modified TLBO was then applied effectively to maximize the cooling rate and COP simultaneously. The determination of the total number of semiconductor elements as well as the supply current to the hot stage and the cold stage were considered as search variables. This research has improved the ability of the TLBO and the modified TLBO as compared to the GA and PSO approaches by evaluating the convergence rate and the computational time. In addition, the obtained results were analyzed using the Wilcoxon signed-rank test. However, the presentation of parameter selection for TLBO and its modified version was not conducted. The main important parameters such as number of generations, population, and teaching factor need to be indicated clearly in the selection of the values used in optimizing the TTEC.

1.3 MATHEMATICAL MODELING OF STEC AND TTEC

Operation of TEC is based on the Peltier effect. TEC acts like a solid-state cooling device that can pump heat from one junction to the other junction when a direct current (DC) is applied (Cheng & Lin, 2005). The energy balance equations at the hot junction and the cold junction for TEC can be described as in Equations 1.1 and 1.2.

$$Q_c = N\left[\alpha I T_c - \frac{1}{2} I^2 \left(\rho_r \frac{L}{A} + \frac{2r_c}{A}\right) - \frac{\kappa A (T_h - T_c)}{L}\right] \tag{1.1}$$

$$Q_h = N\left[\alpha I T_h + \frac{1}{2} I^2 \left(\rho_r \frac{L}{A} + \frac{2r_c}{A}\right) - \frac{\kappa A (T_h - T_c)}{L}\right] \tag{1.2}$$

where Q_c is the cooling rate that will be absorbed from the cold side of TEC and Q_h is the heat rejection that will be released to the environment. Equations 1.1 and 1.2 show the completion between the Seebeck coefficient term, which is responsible for TEC cooling. Besides the parasitic effect of heating, back-heat conduction from the electrical resistance and thermal conductance terms are also represented, respectively. The heat flow $\alpha I T_h$ and $\alpha I T_c$ caused by the Peltier effect are absorbed at the cold junction and released from the hot junction. Joule heating $\frac{1}{2} I^2 \left(\rho_r \frac{L}{A} + 2\frac{r_c}{A}\right)$ due to the flow of electrical current through the material is generated on both sides of the TEC semiconductor couples and at the contact surfaces between the TEC semiconductor couples and the two substrates (Cheng & Lin, 2005). The TEC is operated between temperatures T_c and T_h such that the heat conduction $\kappa A(T_h - T_c)$ occurs through the TEC semiconductor couples. The input electrical power P and COP are calculated using Equations 1.3 and 1.4.

$$P = Q_h - Q_c \tag{1.3}$$

$$\mathrm{COP} = \frac{Q_c}{Q_h - Q_c} \tag{1.4}$$

where α, ρ_r, and κ are the material properties of a thermoelectric couple. Every thermoelectric couple includes one n-type and one p-type semiconductor element which have their own material properties (α_n, ρ_n, κ_n or α_n, ρ_n, κ_n). They represent the thermoelectric material properties. *A*, *L*, and *N* are the cross-sectional area, length, and total number of semiconductor elements, respectively. They represent the geometric properties of the TEC model. COP is a common metric used to quantify the effectiveness of a heat engine. It is also important to quantify the amount of heat that a TEC can transfer and the maximum differential across the TEC. For a STEC, the COP is between 0.3 and 0.7.

As can be seen in Equations 1.1 through 1.4, the important values of the TEC are impacted by three types of parameters. They are operating condition parameters such as temperature difference (ΔT), applied current (I), geometric properties (A, L, N), and finally the material properties (α, ρ_r, κ). Based on the work of Huang et al. (2013), the authors claimed that with a fixed ΔT, cooling rate and COP are first increased and then decreased as the supplied current is increased. Unfortunately, with the same supplied current, a maximum cooling rate and a maximum COP value cannot be always reached simultaneously. Similarly, the optimal cooling rate and COP could not be consistently achieved by varying the TEC geometry.

1.3.1 Impact of Geometric Properties

From Equations 1.1 and 1.2, the maximum cooling rate increases with the decrement of the semiconductor length until it reaches a maximum. Then it decreases with a further reduction in the thermo-element length (Huang et al., 2013). The COP increases with an increase in thermo-element length. As the COP increases with the semiconductor area, the cooling rate may decrease because the total available volume is limited. As the semiconductor area is reduced, the cooling rate generally increases. A smaller semiconductor area and a greater number of them yield greater cooling capacity. When the semiconductor length is below than this lower bound, the cooling capacity declines enormously (Rowe & Min, 1996). Other elements do affect the performance of the TEC (e.g., contact resistance) but since it is very small in most calculations, it can be neglected.

1.3.2 Impact of Material Properties

The performance of TECs strongly depends on thermoelectric materials (Huang et al., 2013). A good thermoelectric material should have a large Seebeck coefficient to get the greatest possible temperature difference per given amount of electrical potential (voltage). In addition, low electrical resistance to minimize the Joule heating (Yamashita & Tomiyoshi, 2004) and low thermal conductivity to reduce the conduction from the hot side and back to the cold side are also crucial. Pure metal has a low Seebeck coefficient which leads to low thermal conductivity, whereas in insulators electrical resistivity is low which leads to higher Joule heating. The performance evaluation index of thermoelectric materials is the figure of merit (Z) or the dimensionless figure of merit ($ZT = \alpha^2 T/\rho\kappa$) which combines the aforementioned properties. The increase in Z or ZT leads directly to the improvement in the cooling efficiencies of Peltier modules. The material properties are considered to be dependent on the average temperature of the cold side and hot side of each stage. Their values can be calculated using Equations 1.5 and 1.6 (Cheng & Lin, 2005):

$$\alpha_p = -\alpha_n = \left(-263,38 + 2.78T_{ave} - 0.00406T_{ave}^2\right) \times 10^{-6}$$

$$\alpha = \alpha_p - \alpha_n; \tag{1.5}$$

$$\rho_p = \rho_n = \left(22,39 - 0.13T_{ave} + 0.00030625T_{ave}^2\right) \times 10^{-6}$$

$$\rho_r = \rho_p + \rho_n; \tag{1.6}$$

For TTEC, there are three types which have three typical approaches to supply the electric current to each stage: in series, in parallel, and separated (Cheng & Shih, 2006). When a STEC is not able to work within the required temperature difference for a specific application, it is possible to use multistage configurations to extend the temperature difference (Enescu & Virjoghe, 2014). In the research of Enescu and Virjoghe (2014), mathematical modeling of TTEC was presented as follows:

$$Q_{c,c} = \frac{N_t}{r+1}\left[\alpha_c I_c T_{c,c} - \frac{1}{2} I_c^2 R_c - K_c\left(T_{c,h} - T_{c,c}\right)\right] \tag{1.7}$$

$$Q_{c,h} = \frac{N_t}{r+1}\left[\alpha_c I_c T_{c,h} + \frac{1}{2} I_c^2 R_c - K_c\left(T_{c,h} - T_{c,c}\right)\right] \tag{1.8}$$

$$Q_{h,c} = \frac{N_t r}{r+1}\left[\alpha_h I_h T_{h,c} - \frac{1}{2} I_h^2 R_h - K_h\left(T_{h,h} - T_{h,c}\right)\right] \tag{1.9}$$

$$Q_{h,h} = \frac{N_t r}{r+1}\left[\alpha_h I_h T_{h,c} + \frac{1}{2} I_h^2 R_h - K_h\left(T_{h,h} - T_{h,c}\right)\right] \tag{1.10}$$

where $Q_{c,c}$ is the cooling capacity at the cold side of the colder stage. $Q_{c,h}$ is the released heat rate at the hot side of the colder stage. $Q_{h,c}$ is the cooling capacity at the cold side of the hotter stage, and $Q_{h,h}$ is the release heat rate at the hot side of the hotter stage. N_t is the total number of semiconductor elements which were put inside TTEC. r is the ratio between the number of semiconductor elements of the hot stage and the number of semiconductor elements of the cold stage. I_c and I_h are the applied current to the colder stage and hotter stage, respectively. $T_{c,c}$ and $T_{c,h}$ represent the cold- and hot-side temperatures of the colder stage. $T_{h,c}$ and $T_{h,h}$ represent the cold- and hot-side temperatures of the hotter stage while α_h, R_h, and K_h are the Seebeck coefficient, the electrical resistance, and the thermal conductance of the colder stage, respectively. α_c, R_c, and K_c are the Seebeck coefficient, the electrical resistance, and the thermal conductance of the colder stage, respectively. The properties of the thermoelectric material of each stage i can be determined as follows:

$$\alpha_i = \left(\alpha_{i,p} - \alpha_{i,n}\right)_{T_{i,\text{ave}}} \tag{1.11}$$

$$R_i = \frac{\left(\rho_{i,p} + \rho_{i,n}\right)_{T_{i,\text{ave}}}}{G} \tag{1.12}$$

$$K_i = \left(\kappa_{i,p} + \kappa_{i,n}\right)_{T_{i,\text{ave}}} G \tag{1.13}$$

The subscripts p and n indicate the properties of the p-type and n-type semiconductors. ρ and κ are the electrical resistivity and thermal conductivity of the material, respectively. The parameter G is the structure parameter of thermocouples which indicated the ratio of cross-sectional area to the length of thermoelectric modules. The material properties are considered to depend on the average temperature $T_{i,\text{ave}}$ of the cold-side temperatures of each stage and their values can be calculated using the following equations (i = c and h) (Venkata Rao & Patel, 2013):

$$\alpha_{i,p} = -\alpha_{i,n} = \left(22224.0 + 930.6T_{i,\text{ave}} - 0.9905T_{i,\text{ave}}^2\right) \times 10^{-9} \tag{1.14}$$

$$\rho_{i,p} = \rho_{i,n} = \left(5112.0 + 163.4T_{i,\text{ave}} + 0.6279T_{i,\text{ave}}^2\right) \times 10^{-10} \tag{1.15}$$

$$k_{i,p} = k_{i,n} = \left(62605.0 - 277.7T_{i,\text{ave}} + 0.4131T_{i,\text{ave}}^2\right) \times 10^{-4} \tag{1.16}$$

Equation 1.17 presents the formulation of the COP of TTEC. Similar to STEC, COP is the ratio between the cooling capacity of the cold side and the electrical power consumption, P. It is also important to quantify the amount of heat that a TEC can transfer and the maximum differential across the TEC.

$$\text{COP} = \frac{Q_{c,c}}{Q_{h,h} - Q_{c,c}} = \frac{Q_{c,c}}{P} \tag{1.17}$$

Constriction and spreading resistances exist whenever heat flows from one region to another of different cross-sectional area. The term "constriction" is used to describe the situation where heat flows into a narrower region while "spreading resistance" describes the case where heat flows out of a narrow region into a larger cross-sectional area (Song, Au, & Moran, 1995). The total thermal resistance RS_t existing between the interfaces of TTEC is taken into consideration with this result and is given in Equation 1.18, where RS_{cont} and RS_{sprd} are contact thermal resistance and spreading thermal resistance between the interfaces of two single stages of TTEC, respectively.

Based on the work of Song et al. (1995) and Venkata Rao and Patel (2013), the RS_{cont} and RS_{sprd} are calculated in Equations 1.19 and 1.20:

$$RS_t = RS_{cont} + RS_{sprd} \tag{1.18}$$

$$RS_{cont} = \frac{RS_j}{\frac{2aN_t}{r+1}} \tag{1.19}$$

$$RS_{sprd} = \frac{\psi_{max}}{k_{h,s} \times rad_{c,s}\sqrt{\pi}} \tag{1.20}$$

In Equation 1.19, RS_j is the joint resistance at the interface of two single stages of TTEC. The factor $2a$ represents the linear relationship between the cross-sectional area of the substrate and the thermo-element modules. In Equation 1.20, $k_{h,s}$ is the thermal conductivity of the substrate of the hot stage and $rad_{c,s}$ is the equilibrium radius of the substrates of the cold stage. Detailed explanations related to $rad_{c,s}$ are available in Song et al. (1995). From there, the calculation can be expressed as follows:

$$rad_{c,s} = \sqrt{\frac{\frac{2aN_t}{r+1}}{\pi}} \tag{1.21}$$

The dimensionless value ψ_{max} of Equation 1.20 is given in Equation 1.22. ε, τ and ϕ are the dimensionless parameters and are calculated in Equations 1.23 through 1.25. $S_{h,s}$ is the substrate thickness of the hot stage and $rad_{h,s}$ is the equilibrium radius of the substrate of the hot stage, respectively. Bi is the Biot number and its value is infinity (Bi = ∞) for the isothermal cold side of the hot stage. The dimensionless parameters λ and $rad_{h,s}$ of Equations 1.26 and 1.27 are given by Cheng and Shih (2006).

$$\psi_{max} = \frac{\varepsilon \times \tau}{\sqrt{\pi}} + \frac{1}{\sqrt{\pi}}(1-\varepsilon)\phi \tag{1.22}$$

$$\varepsilon = \sqrt{\frac{1}{r}} \tag{1.23}$$

$$\tau = \frac{S_{h,s}}{rad_{h,s}} \tag{1.24}$$

$$\phi = \frac{\tanh(\lambda \times \tau) + \frac{\lambda}{Bi}}{1 + \frac{\lambda}{Bi}\tanh(\lambda \times \tau)} = \tanh(\lambda \times \tau) \tag{1.25}$$

$$\text{rad}_{\text{h,s}} = \sqrt{\frac{\frac{2aN_t r}{r+1}}{\pi}} \tag{1.26}$$

$$\lambda = \pi + \frac{1}{\varepsilon\sqrt{\pi}} \tag{1.27}$$

The hot side of the cold stage and cold side of the hot stage are at the interface; so $Q_{\text{h,c}} = Q_{\text{c,h}}$, but due to the thermal resistance at the interface, the temperatures of both sides are not the same. The relationship between both these temperatures is given as follows (Cheng & Shih, 2006):

$$T_{\text{h,c}} = T_{\text{c,h}} + RS_t \times Q_{\text{c,h}} \tag{1.28}$$

Since $Q_{\text{h,c}} = Q_{\text{c,h}}$, we can obtain Equation 1.29 as follows:

$$\frac{N_t}{r+1}\left[\alpha_c I_c T_{\text{c,h}} + \frac{1}{2}I_c^2 R_c - K_c\left(T_{\text{c,h}} - T_{\text{c,c}}\right)\right] = \frac{N_t}{r+1}\left[\alpha_c I_c T_{\text{c,h}} + \frac{1}{2}I_c^2 R_c - K_c\left(T_{\text{c,h}} - T_{\text{c,c}}\right)\right] \tag{1.29}$$

By replacing $T_{\text{h,c}}$ via plugging Equation 1.28 into Equation 1.29, the hot-side temperature of the colder stage can be found as in Equation 1.30:

$$T_{\text{c,h}} = \frac{\left(\frac{1}{2}I_c^2 R_c + K_c T_{\text{c,c}}\right)\left(r\alpha_h I_h RS_t \frac{N_t}{r+1} + rK_h RS_t \frac{N_t}{r+1} - 1\right) - r\left(\frac{1}{2}I_h^2 R_h + K_h T_{\text{h,h}}\right)}{\left(\alpha_c I_c - K_c\right)\left(1 - r\alpha_h I_h RS_t \frac{N_t}{r+1} - rK_h RS_t \frac{N_t}{r+1}\right) - r\left(\alpha_h I_h + K_h\right)} \tag{1.30}$$

1.4 TEC DESIGN FORMULATION

The main objective of the TEC design is to determine a set of design parameters which yields the maximum cooling rate and/or maximum COP while meeting the required specifications at minimal cost. Based on previous works, the main parameters are classified into four groups for the optimization of the design of STEC or TTEC:

1. *Group 1—Objective functions*: The objective function is the maximum cooling rate and/or COP. For STEC, the formulations of cooling rate and COP are presented in Equations 1.1 and 1.4, respectively. For TTEC, the formulation of the cooling rate and COP are presented in Equations 1.7 and 1.17. These objectives can be optimized individually or simultaneously based on SOO or MOO strategies.
2. *Group 2—Variables*: For STEC, the design parameters are:
   a. Length of semiconductor element (L)
   b. Cross-sectional area of semiconductor element (A)
   c. Number of semiconductor elements (N)

For TTEC, the design parameters are:

a. Supplied current to hot stage (I_h) and cold stage (I_c)
b. Ratio of number of semiconductor elements (r) between hot stage and cold stage

3. *Group 3—Fixed parameters*: For STEC, some fixed paramcters need to be determined as follows:
   a. Total volume in which STEC can be placed which is determined by total cross-sectional area (S) and its height (H)
   b. Operating conditions such as applied current I and required temperature at hot side T_h and cold side T_c of STEC. Then, the material properties are calculated based on Equations 1.5 and 1.6.

   For TTEC, the fixed parameters are as follows:
   a. Total number of semiconductor elements (N_t) in both stages
   b. Operating conditions of the system, such as the required temperature at hot side of the hot stage $T_{h,h}$ and cold side of the cold stage $T_{c,c}$
4. *Group 4—Constraints*: For both types of TEC, the constraints are:
   a. Boundary constraints on the design variables
   b. The requirement of satisfying a required value of COP (COP > COP_{min}) and a limited value of the manufactured cost (cost < cost max).

For ease of formulation and implementation, the penalty function approach needs to be used to convert the constrained problem into an unconstrained one to adjust the infeasible search space (Bryan & Shibberu, 2005). Hence, for the SOO problem, the formulation of the objective function is presented in Equation 1.31:

$$\text{maximize } F = \begin{bmatrix} \text{Cooling rate} \\ \\ \text{COP} \end{bmatrix} + \beta \cdot \sum_{i=1}^{n} f_{i,\text{constraint}} \tag{1.31}$$

where F is an objective function, which consists of two parts: first is the objective function cooling rate of COP and the second is the penalty function (which contains the constraint violation function $f_{i,\text{constraint}}$) and the coefficient of violation, $\beta \cdot \beta$ is set by the user and is normally chosen as a large value (e.g., 10^{15}). During the search, $f_{i,\text{constraint}}$ will be aequal to 1 if the search variables do not satisfy one of the constraints, making the function F to blow up to infinity. If all the constraints are satisfied where all $f_{i,\text{constraint}}$ is equal to 0 and maximize F = maximize (cooling rate or COP), the search variables are within feasible ranges.

MOO has been defined as finding a vector of the decision variables while maximizing or minimizing several objectives simultaneously within some constraint conditions (Kim & De Weck, 2006). In MOO, the definition of a performance is more complex than an SOO problem because the optimization goal itself consists of multiple target objectives. Therefore, a single desirable best solution with respect to all the objectives does not exist. However, a series of good solutions which are equally good and known as Pareto optimal solutions could be attained (Babu & Angira, 2003). Multiobjective programming models are difficult to solve because of the incoordination of the target vectors and the existence of constraints. There must be a dimensionless or a dimensionally unified treatment for solving MOO problems whose dimensions are not uniform. In this chapter, two main objective functions of the TEC model are the cooling rate (in Watts) and the COP (dimensionless unit). They need to be optimized simultaneously by combining them into one dimensionless objective function. The weighted sum method or scalarization method is commonly used to solve MOO problems by combining its multiple objectives into one single-objective scalar function (Deb, 2014). As can be seen in Equations 1.32 and 1.33, the weighted sum method is conducted by multiplying each objective function with a weighting factor and summing up all these weighted objective functions.

$$\text{Maximize } F = w_1 \frac{\text{cooling rate}}{(\text{cooling rate})_{\text{max}}} + w_2 \frac{\text{COP}}{\text{COP}_{\text{max}}} + \beta \cdot \sum_{i=1}^{n} f_{i,\text{constraint}} \tag{1.32}$$

$$w_1 + w_2 = 1 \tag{1.33}$$

w_1 and w_2 are the weighting factors for each objective function. The Pareto front is obtained by changing the w_1 and w_2 systemically to create respectively different optimal solutions. In MOO, the image of all optimal solutions is called the Pareto front. The shape of the Pareto front indicates the nature of the trade-off between the different objective functions (Deb, 2014). The Pareto front contains the Pareto optimal set of solutions such that when going from any point to another in the set, at least one objective function improves and at least one other worsens (Babu & Angira, 2003). The Pareto front divides the objective function space into two parts: one part contains nonoptimal solutions and another part contains the infeasible solutions. For nonlinear MOO, determining the entire continuous Pareto-optimal surface is practically impossible. However, finding a discrete set of Pareto-optimal points which approximates the true Pareto front is a realistic expectation (Boussaïd, Lepagnot, & Siarry, 2013).

1.5 DE FOR TEC DESIGN

DE is one of the popular evolutionary metaheuristic algorithms like GA and PSO (Storn & Price, 1995). DE is divided into ten different strategies. A strategy that works out to be the best for a given problem may not work well when applied for a different problem (Babu & Angira, 2003). DE has been used widely to solve problems which are non-differentiable, noncontinuous, nonlinear, noisy, flat, multidimensional, and have many local minima. The optimization search of DE proceeds over three operators: mutation, crossover, and selection (Price, Storn, & Lampinen, 2006). At the first stage, a population of candidate solutions for the optimization task to be solved is randomly initialized (Boussaïd et al., 2013). One target vector X_i^{Pri} is randomly selected in the population. For every generation of the evolution process, new individuals are created by applying reproduction mechanisms which are the crossover and mutation operators. During mutation, the mutant vector V_i is generated by combining three randomly selected vectors X_1^{aux}, X_2^{aux}, X_3^{aux} from the population vector excluding target vector $X_i^{\text{Pr}i}$:

$$V_i = X_1^{\text{aux}} + F(X_2^{\text{aux}} - X_3^{\text{aux}}) \tag{1.34}$$

where F is the mutation amplification factor which is a positive real number that controls the rate at which the population evolves. In the crossover process, DE performs a uniform crossover between the target vector $X_i^{\text{Pr}i}$ and the mutant vector V_i to create a trial vector X_i^{child} (see Equation 1.35). The crossover probability, CR, which is used to control the fraction parameter values that are copied from the mutant vector, must be specified within the range [0,1]. To determine which source contributes a given vector, uniform crossover compares CR to the output of a uniform random number generator, rand(0,1). If the random number is less than or equal to CR, the trial vector is inherited from the mutant vector. Otherwise, the vector is copied from the target vector:

$$X_i^{\text{child}} = \begin{cases} V_i & \text{if}\,(\text{rand}(0,1) \le \text{CR}) \\ X_i^{\text{Pr}i} & \text{if}\,(\text{rand}(0,1) > \text{CR}) \end{cases} \tag{1.35}$$

Finally, the fitness of the resulting solutions is evaluated in the selection process and the target vector of the population competes against a trial vector to determine which one will be retained into the next generation (shown in Equation 1.36). The executional procedures of DE are the same as with GA except for the differences in the mutation type and reproduction mechanisms:

$$X_{i+1}^{\text{Pr}i} = \begin{cases} X_i^{\text{child}} & \text{if } f\left(X_i^{\text{child}}\right) < f\left(X_i^{\text{Pr}i}\right) \\ X_i^{\text{Pr}i} & \text{otherwise} \end{cases} \tag{1.36}$$

Parameter selection of DE is shown in Table 1.2 using deterministic rules.

DE is highly sensitive to the choice of scaling factor F. The bigger the value of F, the higher the exploration capability of DE (Guo et al., 2014). A good initial guess is to choose F within the range of [0.5, 1]; for example, $F = 0.85$ would be a good

TABLE 1.2
Parameter Settings of the DE Algorithm

Parameters of DE	Specified Value
Number of population members	$P = 30$
Scaling factor	$F = 0.85$
Crossover probability constant	$\text{CR} = 1$
Number of variables	$D = 3$
Maximum number of iterations	$i_{\max} = 300$

initial choice (Storn & Price, 1997). It is said that the value of F smaller than 0.4 and greater than 1 is occasionally effective. To choose the suitable value for crossover probability CR, a bigger CR can increase the convergence speed of the algorithm but a smaller CR could increase the exploitation capability (Guo et al., 2014). The value of CR is chosen within the range of [0,1] to help maintain the diversity of the population. However, for most cases, it should be close to 1 (e.g., CR > 0.9) (Storn & Price, 1997). When CR is equal to 1, the number of trial solutions will be reduced dramatically. This may lead to search stagnation. Only separable problems do better with CR close to 0 as [0, 0.2] (Price et al., 2006). Choosing values for the number of population members, P, is not very critical. An initial guess (10D) is a good choice to obtain global optimum (Guo et al., 2014), where D stands for a number of variables. Depending on the difficulty of the problem, the number of the population, P, can be lower than (10D) or higher than it to achieve convergence such as 5D to 10D, 3D to 8D, or 2D to 40D (Storn & Price, 1997). In the stopping condition, the algorithm will stop if number of function evaluations exceeds its maximum value (e.g., $i_{max} = 300$). The DE algorithm (Ganesan, Elamvazuthi, Shaari, & Vasant, 2014) is described as follows while the flowchart of the DE algorithm is shown in Figure 1.3.

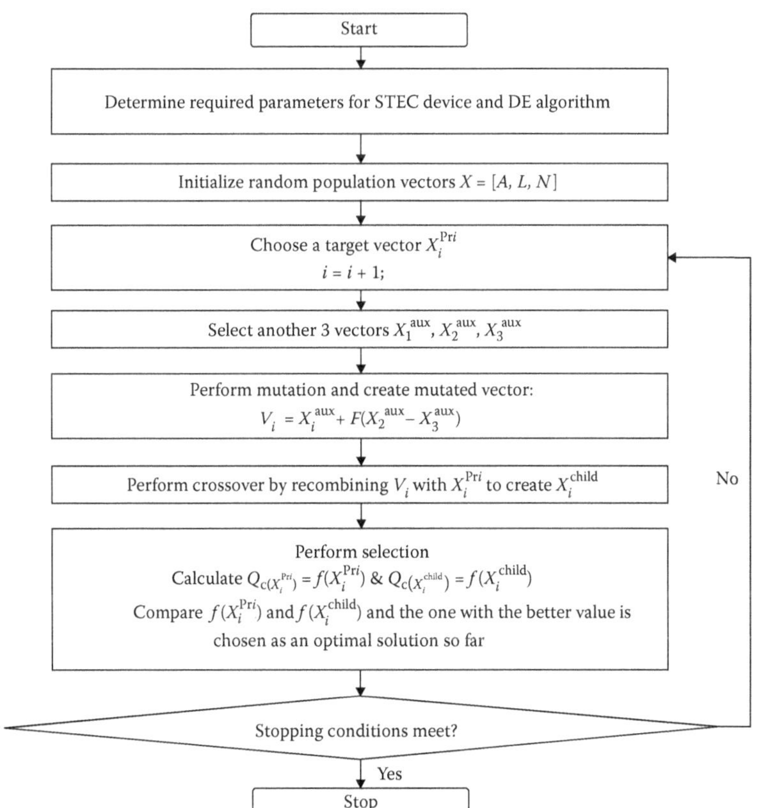

FIGURE 1.3 Flowchart of DE algorithm with TEC model.

- *Step 1*: Initial parameter setting for STEC model and DE algorithm. For DE, set required parameters for the algorithm as in Table 1.2. For a TEC device, set required parameters such as fixed parameters and boundary constraints of the design variables; set all the constraints and apply them into the penalty function.
- *Step 2*: Randomly initialize the population vectors $X_0 = [A, L, N]$ for STEC or $X_0 = [I_h, I_c, r]$ for TTEC within the range of boundary constraints by using generated random numbers method.
- *Step 3*: Randomly select one target vector that can be called as principal parent $X_i^{\mathrm{Pr}i}$ and start the counter $i = i + 1$.
- *Step 4*: Randomly select another three vectors that can be called as auxiliary parents $X_1^{\mathrm{aux}}, X_2^{\mathrm{aux}}$, and X_3^{aux} from the population vectors.
- *Step 5*: Perform differential mutation and generate mutated vector V_i. The mutation vector V_i is defined as $V_i = X_i^{\mathrm{aux}} + F\left(X_2^{\mathrm{aux}} - X_3^{\mathrm{aux}}\right)$.
- *Step 6*: Perform crossover operation by recombining V_i with $X_i^{\mathrm{Pr}i}$ to generate a trial vector X_i^{child}.
- *Step 7*: Perform selection by comparing the fitness value to choose the suitable vector among the trial vector X_i^{child} and target vector $X_i^{\mathrm{Pr}i}$.
- *Step 8*: If the fitness criterion is satisfied and the number of iterations $i \geq i_{\max}$, stop the algorithm and print optimal solution, else continue to step 3.

1.6 SA FOR TEC DESIGN

SA is a method employed for solving both unconstrained and constrained optimization problems (Blum & Roli, 2008). The method is based on the models related to the metallurgical process of heating a metal/material and then slowly lowering the temperature to decrease undesired properties. During this process, the system's energy is minimized. The objective function of the problem similar to the energy of a material is then minimized by introducing a fictitious temperature which is a simple controllable parameter in the algorithm.

At each iteration of the SA algorithm, a new point x^{k+1} is randomly generated within the boundary constraints based on the current point x^k. The distance of the new point from the current point, or the extent of the search is based on a probability distribution with a scale proportional to the temperature. The algorithm accepts all new points that lower the objective with a certain probability. Points that raise the objective values are evaluated based on the Metropolis calculation (Equation 1.38).

$$f\left(x^{k+1}\right) \leq f\left(x^k\right) \tag{1.37}$$

$$\exp\left(-\frac{f\left(x^{k+1} - x^k\right)}{k_B \cdot T_n}\right) > \mathrm{rand}(0,1) \tag{1.38}$$

where T_n is the current annealing temperature and k_B is Boltzmann annealing. By accepting points that raise the objective, the algorithm avoids being trapped in the

TABLE 1.3
Parameter Settings of SA Algorithm

No.	Parameter Settings	Specific Values
1	Initial temperature	$T_0 = 100$
2	Maximum number of runs	$\text{run}_{\text{max}} = 250$
3	Maximum number of acceptance	$\text{acc}_{\text{max}} = 125$
4	Maximum number of rejections	$\text{rej}_{\text{max}} = 125$
5	Temperature reduction value	$\alpha = 0.95$
6	Boltzmann annealing	$k_B = 1$
7	Stopping criteria	$T_{\text{final}} = 10^{-10}$

local minima and is thus able to explore globally for more possible solutions. An annealing schedule is selected to systematically decrease the temperature as the algorithm proceeds. As the temperature decreases, the algorithm reduces the extent of its search to converge to a minimum.

A programmed SA code was used and its parameters were adjusted so that it could be utilized for finding the optimal TEC design. Choosing good algorithm parameters is very important because it greatly affects the whole optimization process. Parameter settings of SA are listed in Table 1.3.

The initial temperature, $T_0 = 100$, should be high enough such that in the first iteration of the algorithm, the probability of accepting a worse solution, is at least 80%. The temperature is the controlled parameter in SA and it is decreased gradually as the algorithm proceeds (Vasant & Barsoum, 2009). Temperature reduction value $\alpha = 0.95$ and temperature decrease function is:

$$T_n = \alpha T_{n-1} \tag{1.39}$$

The numerical experimentation was done with different α values: 0.70, 0.75, 0.85, 0.90, and 0.95 (Abbasi, Niaki, Khalife, & Faize, 2011). Boltzmann annealing factor, k_B, is used in the Metropolis algorithm to calculate the acceptance probability of the points. Maximum number of runs, $\text{run}_{\text{max}} = 250$, determines the length of each temperature level $T \cdot \text{acc}_{\text{max}} = 125$ determines the maximum number of acceptance of a new solution point and $\text{rej}_{\text{max}} = 125$ determines the maximum number of rejection of a new solution point ($\text{run}_{\text{max}} = \text{acc}_{\text{max}} + \text{rej}_{\text{max}}$) (Abbasi et al., 2011). The stopping criteria determine when the algorithm reaches the desired energy level. The desired or final stopping temperature is set as $T_{\text{final}} = 10^{-10}$. The SA algorithm is described in the following section and the flowchart of SA algorithm is shown in Figure 1.4.

- *Step 1*: Set the initial parameters and create initial point of the design variables. For SA algorithm, determine required parameters for the algorithm as in Table 1.3. For TEC device, set required parameters such as fixed parameters and boundary constraints of the design variables, and set all the constraints and apply them into penalty function.

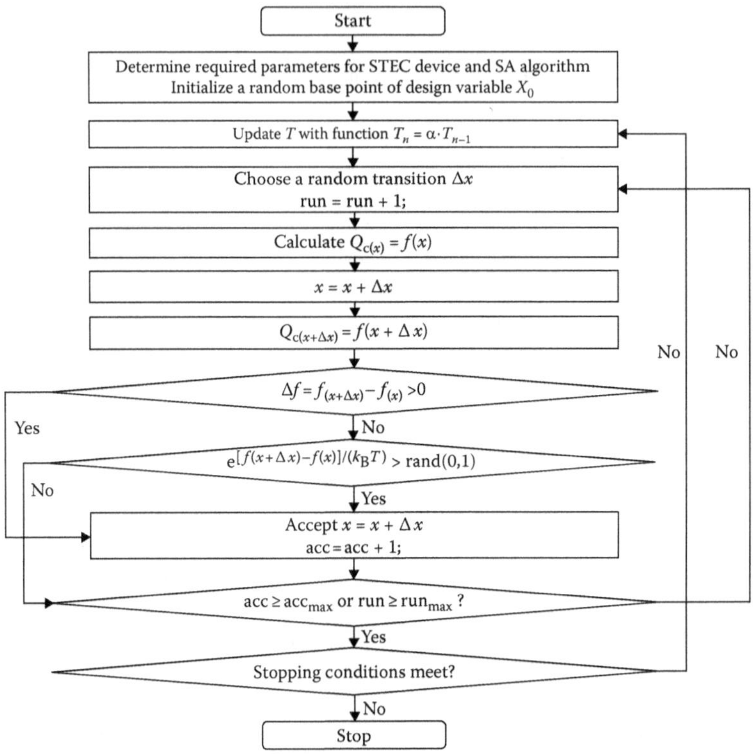

FIGURE 1.4 Flowchart of SA algorithm with TEC model.

- *Step 2*: $X_0 = [A_0, L_0, N_0]$ for STEC or $[I_{h0}, I_{c0}, r_0]$ for TTEC—Initial randomly based point of design parameters within the boundary constraint by computer-generated random numbers method. Then, consider its fitness value as the best fitness so far.
- *Step 3*: Choose a random transition Δx and run = run + 1.
- *Step 4*: Calculate the function value before transition $Q_{c(x)} = f(x)$.
- *Step 5*: Make the transition as $x = x + \Delta x$ within the range of boundary constraints.
- *Step 6*: Calculate the function value after transition $Q_{c(x+\Delta x)} = f(x + \Delta x)$.
- *Step 7*: If $\Delta f = f(x + \Delta x) - f(x) > 0$ then accept the state $x = x + \Delta x$.
- *Step 8*: Else if $\Delta f = f(x + \Delta x) - f(x) \leq 0$, then generate a random number rand in range (0, 1). If $e^{[f(x+\Delta x) - f(x)]/k_B \cdot T} > \text{rand}(0,1)$, then accept the state $x = x + \Delta x$ and acc = acc + 1. Else return to the previous state $x = x - \Delta x$ and rej = rej + 1.
- *Step 9*: If $\text{acc} \geq \text{acc}_{max}$ or $\text{run} \geq \text{run}_{max}$ then continue to step 9. If not, return to step 2.
- *Step 10*: If the process meets the stopping conditions, stop running the SA algorithm, get the optimal value x_{best} and $f(x_{best})$. Otherwise, update T based on temperature reduction function $T_n = \alpha \cdot T_{n-1}$ and return to step 2.

1.7 SUMMARY

In this chapter, the SA and DE algorithms were presented and their application in two types of TEC models (single stage and two stage) was discussed. In the single-stage TEC, the design factors are the geometric parameters of the semiconductor element which are the dimensions of the semiconductor column and the total number of semiconductor units. In the two-stage TEC, the design parameters are the supplied current in the hot and cold stages along with the number of total semiconductor units. The cooling rate and the COP are the two important criteria employed to evaluate the performance of the STEC. They are considered as objective functions and are optimized separately or simultaneously by using a weighted method. Description of the DE and SA algorithms are presented together with the combination of TEC mathematical modelling. Parameter settings of the algorithms are referred from previous studies in combination with the author's experience when doing programming in MATLAB. For future research, further simulation work could be conducted in MATLAB to get the results using metaheuristics. Different test cases could be executed in a systematic way as described in this chapter to evaluate the performance of every algorithm. The comparison between the practical experiments and simulation works would be necessary to analyze and justify the research claims. Developing a prototype for the optimized TEC design based on the optimization procedures highlighted in this chapter would be the ultimate goal of these research efforts.

REFERENCES

Abbasi, B., Niaki, S. T. A., Khalife, M. A., & Faize, Y. (2011). A hybrid variable neighborhood search and simulated annealing algorithm to estimate the three parameters of the Weibull distribution. *Expert Systems with Applications*, *38*(1), 700–708.

Ali, M., Pant, M., & Abraham, A. (2009). Simplex differential evolution. *Acta Polytechnica Hungarica*, *6*(5), 95–115.

Babu, B., & Angira, R. (2003). New strategies of differential evolution for optimization of extraction process. *International Symposium & 56th Annual Session of IIChE* (CHEMCON-2003), Bhubaneswar, India.

Blum, C., & Roli, A. (2008). Hybrid metaheuristics: An introduction. *Hybrid Metaheuristics* (pp. 1–30). Berlin, Heidelberg: Springer.

Boussaïd, I., Lepagnot, J., & Siarry, P. (2013). A survey on optimization metaheuristics. *Information Sciences*, *237*, 82–117.

Bryan, K., & Shibberu, Y. (2005). Penalty functions and constrained optimization. Department of Mathematics, Rose-Hulman Institute of Technology. http://www.rose-hulman.edu/~bryan/lottamath/penalty.pdf.

Cheng, Y.-H., & Lin, W.-K. (2005). Geometric optimization of thermoelectric coolers in a confined volume using genetic algorithms. *Applied Thermal Engineering*, *25*(17), 2983–2997.

Cheng, Y.-H., & Shih, C. (2006). Maximizing the cooling capacity and COP of two-stage thermoelectric coolers through genetic algorithm. *Applied Thermal Engineering*, *26*(8), 937–947.

Davis, L. (1991). *Handbook of genetic algorithms* (Vol. 115). New York: Van Nostrand Reinhold.

Deb, K. (2001). Multi-objective optimization. *Multi-objective optimization using evolutionary algorithms* (pp. 13–46). New York: John Wiley & Sons, Inc.

Deb, K. (2014). Multi-objective optimization. *Search methodologies* (pp. 403–449). New York: Springer.

Deb, K., Pratap, A., Agarwal, S., & Meyarivan, T. (2002). A fast and elitist multiobjective genetic algorithm: NSGA-II. *Evolutionary Computation, IEEE Transactions*, *6*(2), 182–197.

Enescu, D., & Virjoghe, E. O. (2014). A review on thermoelectric cooling parameters and performance. *Renewable and Sustainable Energy Reviews*, *38*, 903–916.

Ganesan, T., Elamvazuthi, I., Shaari, K. Z. K., & Vasant, P. (2014). Hopfield differential evolution for multi-objective optimization of a cement-bonded sand mold system. *International Journal of Management Science and Engineering Management*, *9*(1), 40–47.

Goldsmid, H. J. (2009). The thermoelectric and related effects. *Introduction to thermoelectricity* (pp. 1–6). Berlin, Heidelberg: Springer-Verlag.

Guo, H., Li, Y., Li, J., Sun, H., Wang, D., & Chen, X. (2014). Differential evolution improved with self-adaptive control parameters based on simulated annealing. *Swarm and Evolutionary Computation*, *19*, 52–67. doi: http://dx.doi.org/10.1016/j.swevo.2014.07.001.

Huang, Y.-X., Wang, X.-D., Cheng, C.-H., & Lin, D. T.-W. (2013). Geometry optimization of thermoelectric coolers using simplified conjugate-gradient method. *Energy*, *59*, 689–697.

Karimi, G., Culham, J. R., & Kazerouni, V. (2011). Performance analysis of multi-stage thermoelectric coolers. *International Journal of Refrigeration*, *34*(8), 2129–2135. doi: http://dx.doi.org/10.1016/j.ijrefrig.2011.05.015.

Kim, I. Y., & De Weck, O. (2006). Adaptive weighted sum method for multiobjective optimization: A new method for Pareto front generation. *Structural and Multidisciplinary Optimization*, *31*(2), 105–116.

Mezura-Montes, E., Velázquez-Reyes, J., & Coello Coello, C. A. (2006). A comparative study of differential evolution variants for global optimization. *Proceedings of the 8th Annual Conference on Genetic and Evolutionary Computation* (pp. 485–492). New York: ACM.

Nain, P., Sharma, S., & Giri, J. (2010a). Non-dimensional multi-objective performance optimization of single stage thermoelectric cooler. *Simulated evolution and learning* (pp. 404–413). Berlin, Heidelberg: Springer.

Nain, P. K. S., Giri, J. M., Sharma, S., & Deb, K. (2010b). Multi-objective performance optimization of thermo-electric coolers using dimensional structural parameters. In B. Panigrahi, S. Das, P. Suganthan, & S. Dash (Eds.), *Swarm, evolutionary, and memetic computing* (Vol. 6466, pp. 607–614). Berlin, Heidelberg: Springer.

Poudel, B., Hao, Q., Ma, Y., Lan, Y., Minnich, A., Yu, B., et al. (2008). High-thermoelectric performance of nanostructured bismuth antimony telluride bulk alloys. *Science*, *320*(5876), 634–638.

Price, K., Storn, R. M., & Lampinen, J. A. (2006). *Differential evolution: A practical approach to global optimization*. Berlin, Heidelberg: Springer-Verlag.

Rodgers, P. (2008). Nanomaterials: Silicon goes thermoelectric. *Nature Nanotechnology*, *3*(2), 76.

Rowe, D. (1995). *Thermoelectric handbook* (p. 407). Boca Raton, FL: CRC.

Rowe, D., & Min, G. (1996). Design theory of thermoelectric modules for electrical power generation. *IEE Proceedings-Science, Measurement and Technology*, *143*(6), 351–356.

Rutenbar, R. A. (1989). Simulated annealing algorithms: An overview. *Circuits and Devices Magazine, IEEE*, *5*(1), 19–26.

Scherbatskoy, S. A. (1982). Systems, apparatus and methods for measuring while drilling. Google Patents, US 4520468 A.

Song, S., Au, V., & Moran, K. P. (1995). Constriction/spreading resistance model for electronics packaging. *Proceedings of the 4th ASME/JSME Thermal Engineering Joint Conference.*

Storn, R., & Price, K. (1995). *Differential evolution—A simple and efficient adaptive scheme for global optimization over continuous spaces.* ICSI Berkeley.

Storn, R., & Price, K. (1997). Differential evolution—A simple and efficient heuristic for global optimization over continuous spaces. *Journal of Global Optimization, 11*(4), 341–359.

Van Laarhoven, P. J., & Aarts, E. H. (1987). *Simulated annealing.* The Netherlands: Springer.

Vasant, P. (2013). Hybrid LS–SA–PS methods for solving fuzzy non-linear programming problems. *Mathematical and Computer Modelling, 57*(1), 180–188.

Vasant, P., & Barsoum, N. (2009). Hybrid simulated annealing and genetic algorithms for industrial production management problems. *AIP Conference Proceedings, 1159*, 254–261. doi: http://dx.doi.org/10.1063/1.3223938.

Venkata Rao, R., & Patel, V. (2013). Multi-objective optimization of two stage thermoelectric cooler using a modified teaching–learning-based optimization algorithm. *Engineering Applications of Artificial Intelligence, 26*(1), 430–445. doi: http://dx.doi.org/10.1016/j.engappai.2012.02.016.

Wah, B. W., & Chen, Y. X. (2000). Optimal anytime constrained simulated annealing for constrained global optimization. *Principles and practice of constraint programming–CP 2000* (pp. 425–440). Berlin, Heidelberg: Springer.

Xuan, X. C., Ng, K. C., Yap, C., & Chua, H. T. (2002). Optimization of two-stage thermoelectric coolers with two design configurations. *Energy Conversion and Management, 43*(15), 2041–2052. doi: http://dx.doi.org/10.1016/S0196-8904(01)00153-4.

Yamashita, O., & Sugihara, S. (2005). High-performance bismuth-telluride compounds with highly stable thermoelectric figure of merit. *Journal of Materials Science, 40*(24), 6439–6444.

Yamashita, O., & Tomiyoshi, S. (2004). Effect of annealing on thermoelectric properties of bismuth telluride compounds doped with various additives. *Journal of Applied Physics, 95*(1), 161–169.

Zhang, W., & Hansen, K. M. (2009). An evaluation of the NSGA-II and MOCell genetic algorithms for self-management planning in a pervasive service middleware. *Engineering of Complex Computer Systems, 2009 14th IEEE International Conference* (pp. 192–201). Piscataway, NJ: IEEE.

2 Mean-Variance Mapping Optimization for Economic Dispatch

Economic dispatch (ED) is an essential optimization problem in power generation systems. This problem is often related to fuel cost saving. Real-world ED problems have nonconvex objective functions with complex constraints. This is because of the effects of valve-point loadings, usage of multiple fuel (MF) options, and the existence of prohibited operating zones (POZs). This leads to difficulty in finding the global optimal solution. Over the past decades, various optimization techniques have been applied to ED problems. In general, these techniques can be classified into three main categories. The first category involves methods based on mathematical programming (classical calculus-based techniques). The second and third categories are based on artificial intelligence and hybrid methods, respectively (shown in Figure 2.1). However, nonconvex optimization problems are still a challenge for engineers and decision makers in the industry. Hence, there is always a need for developing new techniques for solving nonconvex problems.

2.1 CLASSICAL TECHNIQUES FOR ED

Traditionally, the objective function of the ED problem is presented using a quadratic approximation and this problem is usually solved via mathematical programming methods. In Wollenberg and Wood (1996), the authors introduced mathematical optimization methods which are: lambda iteration method, Newton's method, gradient search, and dynamic programming. These methods were then applied to solve the ED problem. More techniques were also proposed for solving the ED problem such as linear programming (Parikh & Chattopadhyay, 1996), nonlinear programming (Nanda, Hari, & Kothari, 1994), quadratic programming (Fan & Zhang, 1998), and Maclaurin series-based Lagrangian (MSL) method (Hemamalini & Simon, 2009). Among the previously mentioned techniques, the linear programming methods generate reliable solutions with fast computational time. However, they suffer disadvantages associated with the piecewise linear cost approximation. As for the nonlinear programming methods, they on the other hand suffer from problems related to convergence and algorithm complexity. The Newton-based algorithms have difficulty in handling a large number of inequality constraints (Al-Sumait, Al-Othman, & Sykulski, 2007). The MSL method can directly deal with the nonconvex ED problem by using the Maclaurin expansion of the nonconvex terms in the objective function. Although this method can quickly find a solution for the problem, the obtained solution quality is not high, especially

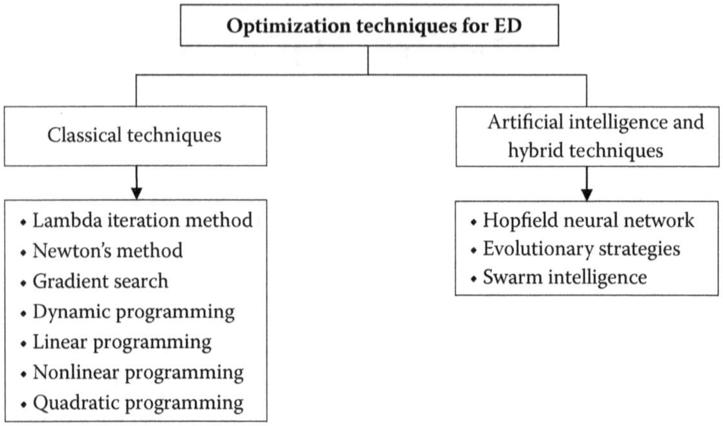

FIGURE 2.1 Optimization techniques for ED.

for large-scale systems. In general, the conventional methods are not capable of solving nonconvex ED problems (Dieu, Schegner, & Ongsakul, 2013).

2.2 ED USING ARTIFICIAL INTELLIGENCE TECHNIQUES

The quadratic cost function is not an accurate representation of the practical ED problem. This is due to the fact that the practical ED problem contains various nonconvex elements and nonlinearities in the objective function and constraints. The effects of valve-point loadings, usage of MF options, and the existence of POZs can cause the input–output curve of thermal generators to become more complicated. For this reason, the practical ED problem should be formulated with a nonconvex objective function. More advanced methods based on artificial intelligence have been previously developed to deal with such ED problems (e.g., Hopfield neural network [HNN], evolutionary programming [EP], differential evolution [DE], genetic algorithm [GA], ant colony optimization [ACO], bacteria foraging optimization [BFO], artificial bee colony [ABC] algorithm, and particle swarm optimization [PSO]).

2.2.1 Hopfield Neural Network

HNNs are a form of recurrent artificial neural network discovered in the 1980s (Park, Kim, Eom, & Lee, 1993). The HNN method is based on the minimization of its energy function. Thus, it is very suitable for implementation in optimization problems.

In Park et al. (1993), the authors formulated the ED problem with piecewise quadratic cost functions by using the HNN. The results obtained using this method were then compared with those obtained using the hierarchical approach. However, the implementation of the HNN to this problem involved a large number of iterations and often produced oscillations (Lee, Sode-Yome, & Park, 1998). In

Lee et al. (1998), a unified learning approach was developed by embedding the slope adjustment and bias adjustment methods into the HNN to speed up its convergence when solving the ED problem.

The improvements of the HNN have been developed to overcome its drawbacks during implementation. These improvements provide the HNN with the ability to find near-optimal solutions for nonconvex optimization problems. The typical improved HNN is the augmented Lagrange HNN which has been applied in complex ED problems which involve MF options and the existence of prohibited zones. However, the unsuitable sigmoid function adopted in the Hopfield model may cause large number of iterations during execution which results in high computational time. This method can be easily implemented on large-scale systems but it suffers from lengthy computational time and stagnation at the local optima (Dieu et al., 2013).

2.2.2 Evolutionary Strategies

Evolutionary strategies are approaches that spring from evolutionary concepts, for instance, EP, DE, and GA. EP is a stochastic search technique which places emphasis on the behavioral linkage between parents and their offspring (Rahimullah, Ramlan, & Rahman, 2003). It is based on three operations: selection, mutation, and competition. Since 2003 modifications and improvements have been proposed and investigated on a variety of fields in engineering. Various EP techniques have been used to solve complex ED problems. In Sinha, Chakrabarti, and Chattopadhyay (2003), four EP techniques including classical EP (CEP), fast EP (FEP), improved FEP (IFEP), and FEP using the weighted mean of Gaussian and Cauchy mutations (MFEP) were applied. These techniques were developed with adaptations based on a scaled cost function and empirical learning rates. Their performances were examined on ED problems with nonconvex cost curves and different system size. These techniques prove to be very effective in solving the nonlinear ED problem without any restrictions on the shape of the cost curves. However, solutions obtained using the EP-based method may get trapped in suboptimal zones when faced with large-scale problems. In general, these methods have a high computaiontal cost. Besides, their optimal solutions are very sensitive to the related controlled parameters (Park, Lee, Shin, & Lee, 2005).

Another popular evolutionary strategy for solving ED problem is the DE approach. DE is a population-based stochastic search technique introduced by Storn and Price in the 1990s (Storn & Price, 1997). DE has many advantages due to its simple/compact structure, few control parameters, and high convergence characteristics. However, there is no guarantee that this method will always obtain the optimal solution. In Noman and Iba (2008), the authors used a classical DE algorithm to solve ED problems with valve-point effects and MF options while considering transmission loss, ramp-rate limits, and the existence of POZs. The DE technique found better solutions as compared to the previous methods in terms of cost, power loss, and mean performance. In all these experiments, the average performance of DE was outstanding and the required time was very much suitable for online solving. However, DE method suffers from high computational time when dealing with large-scale problems.

Prior to DE, GA was one of the most used evolutionary approaches. GA became popular through the work of John Holland in the early 1970s (Holland, 1975). The initial population of GA consists of three main genetic operations: selection, crossover, and mutation. GA is a global and parallel search technique that can handle optimization problems with nonconvex objective functions. GA has been successfully used to solve many ED problems in the past. Among the main drawbacks of GA are long computation time and non-convergence to a global solution. Over the years, there are a lot of enhancements and combinations of GA with other artificial techniques to overcome its drawbacks. In Chiang (2005), the author presented solutions for the ED problem (with valve-point effects and MF options) using the improved GA with multiplier update (IGA_MU). This algorithm is the integration of the Improved GA (IGA) and the multiplier updating (MU) approach. Comparing the results between the previous methods and the conventional GA with the MU (CGA_MU), the proposed algorithm (IGA_MU) was found to be more effective.

2.2.3 Swarm Intelligence

ACO is among the most effective swarm intelligence-based algorithms (Pothiya, Ngamroo, & Kongprawechnon, 2010). The original idea was based on the behavior of ants seeking the shortest path between their colony and food sources. The ACO algorithm consists of four stages: solution construction, pheromone update, local search (LS), and pheromone re-initialization (Pothiya et al., 2010). The ACO algorithm has been implemented as a solution method for ED problems. In Pothiya et al. (2010), ACO was used for solving ED problems with nonsmooth cost functions while taking into account valve-point effects and MF options. To improve the search process, three techniques including the priority list method, variable reduction method, and the zoom feature method were added to the conventional ACO. The near-optimal solutions acquired from the results signify that the ACO provides better solutions as compared to other methods. ACO converges to the optimum solution much faster than the other methods (PSO, TS, GA) employed in Pothiya et al. (2010).

Similar to ACO, BFO is a swarm-based optimization technique that uses population search and global search methods (Padmanabhan, Sivakumar, Jasper, & Victoire, 2011). The BFO uses ideas from natural evolution for efficient search operations. The law of evolution states that organisms with better foraging strategies would survive while those with poor foraging strategies would be eliminated. The foraging behavior of *Escherichia coli* (*E. coli*) bacteria is adopted in the BFO technique. This foraging strategy consists of four processes, namely: chemotaxis, swarming, reproduction, and elimination/dispersal (Padmanabhan et al., 2011). In Padmanabhan et al. (2011), the BFO algorithm was proposed for solving the nonconvex ED. The proposed method was tested on two power systems consisting of 6 and 13 thermal units while considering valve-point effects. The obtained results show that the proposed method had better solution quality, convergence characteristics, computational efficiency, and robustness as compared to other methods.

The ABC algorithm proposed by Karaboga in 2005 is a population-based optimization tool (Karaboga, 2005). The core concept of the ABC algorithm involves the foraging behavior of three types of bees in the honeybee colonies (employed bees, onlooker bees, and scout bees). Each type of bee has different responsibilities in the colony. The employed bees give information to the onlooker bees about the food sources which they found by swarming. The onlooker bees watch all dances of employed bees and assess the food sources. Then they select one of them for foraging. When a food source is abandoned, some employed bees turn to scout bees. The scout bees search for new food sources in the environment. In the ABC algorithm, the location of a food source indicates a potential solution while the nectar amount in the food source refers to the fitness value (Aydin & Özyön, 2013). In Hemamalini and Simon (2010), the ABC algorithm was proposed for solving the nonconvex ED problem which considers valve-point effects, MF options, existence of POZs, and ramp-rate limits. The proposed algorithm was tested on the cases consisting of 10, 13, 15, and 40 generating units with nonsmooth cost functions. The comparison of the results with other methods reported in Hemamalini and Simon (2010) proves the superiority of the proposed method. The method is simple, easy to implement, and has a good convergence rate. In Aydin and Özyön (2013), the authors proposed the incremental artificial bee colony approach (IABC) and incremental artificial bee colony with LS technique (IABC-LS). These approaches were used for solving the ED problem with valve-point effects. The proposed methods were applied to systems with 3, 5, 6, and 40 generators. The results of the algorithms were compared with several other approaches in that work. The obtained results using the proposed methods were seen to be better than the results produced by the other approaches.

In the 1990s, the PSO technique was becoming popular in various fields of study (Mahor, Prasad, & Rangnekar, 2009). PSO is a population-based stochastic search optimization technique motivated by the social behavior of fish schooling and birds flocking. The PSO algorithm searches in parallel using a swarm consisting of a number of particles to explore optimal regions. In PSO, each particle's position represents an individual potential solution to the optimization problem. Each particle's position and velocity are randomly initialized in the search space. Each particle then swarms around in a multidimensional search space directed by its own experience and the experience of neighboring particles. PSO can be applied to global optimization problems with nonconvex or nonsmooth objective functions. Recently, PSO is the most post popular method applied for solving ED problems. Several inproved PSO methods and their hybrids have been developed and proposed for solving nonconvex ED problems.

In Park et al. (2005), the authors proposed a modified PSO (MPSO) approach to deal with the equality and inequality constraints of the ED problems. In that work, a dynamic search-space reduction strategy was devised to accelerate the optimization process. This approach was implemented to ED problems with smooth cost functions and nonsmooth cost functions which consider valve-point effects and MF options. The results of the MPSO were compared with the results obtained using the Tabu Search, EP, GA, and modified HNN. The comparisons indicated that the MPSO has a higher probability of producing the global solution for such ED problems as compared to other heuristic approaches.

In Selvakumar and Thanushkodi (2007), the authors proposed a new version of the classical particle swarm with a split-up in the cognitive behavior. This version was used to solve the nonconvex ED problem. This modification helps to explore the search space very effectively. A simple local random search (LRS) procedure was integrated with the NPSO in order to effectively exploit the promising solution regions. The proposed NPSO–LRS method was applied to three test systems having nonconvex solution spaces. Significantly better results were obtained by the NPSO as compared to the results obtained from previous approaches.

In Chaturvedi, Pandit, and Srivastava (2008), a novel Self-Organizing Hierarchical Particle Swarm Optimization (SOH_PSO) was proposed for the nonconvex ED problem to handle the problem of premature convergence. In the SOH_PSO, the particle velocities were reinitialized whenever the population stagnated at the local optima during search operations. This proposed method was compared with other forms of PSO algorithms. The study concluded that the SOH_PSO is very much capable of achieving global solutions due to its simple structure and its dynamic convergence characteristics.

In Kuo (2008), the author proposed an approach and a coding scheme for solving ED problems using the hybrid simulated annealing-PSO (SA-PSO) technique. This coding scheme could effectively prevent the algorithm from obtaining nonfeasible solutions through the application of stochastic search methods. This approach dramatically improves the search efficiency and the solution quality. In that work, the cost function included valve-point effects along with the operational constraints (such as generation limitations, ramp-rate limits, existence of POZs, and transmission loss). The proposed method was demonstrated by four system case studies and compared with previous works in terms of solution quality and computational efficiency. The encouraging simulation results showed that the proposed method is capable of obtaining more efficient and higher-quality solutions for ED problems.

In Dieu, Schegner, and Ongsakul (2011), a newly improved particle swarm optimization (NIPSO) was proposed for solving the ED problem with valve-point loading effect (VPE). The proposed NIPSO is based on the PSO with time-varying acceleration coefficient (PSO-TVAC). The NIPSO was enhanced by including the use of the sigmoid function with random variation for inertial weight factor. Besides, the pseudo-gradient was employed for improved particle guidance and a quadratic programming scheme was utilized for obtaining the initial conditions. The proposed NIPSO which was tested on 13- and 40-unit systems obtained better optimal solution than most of the other methods.

In Abedinia, Amjady, Ghasemi, and Hejrati (2013), a hybrid PSO with time-varying acceleration coefficients and bacteria foraging algorithm (HPSOTVAC/BFA) was proposed for solving a complex ED problem. Many realistic constraints including generation limitation, ramp-rate limits, existence of POZ, and transmission loss were considered in that work. The effectiveness of the proposed HPSOTVAC/BFA was tested in 6-, 15-, and 40-unit generating systems. The numerical results demonstrated that the proposed hybrid technique gives less total generation costs than the GA, PSO, hybrid GA-PSO, active power optimization, Iterative PSO, and SOH_PSO methods.

Although these PSO-based methods can provide high-quality solutions with short computational time and stable convergence property, their performances seem to be highly dependent on the tuning of the weights and algorithmic parameters. Over the years, the PSO method has been continuously improved for dealing with large-scale and complex problems in power systems. Metaheuristic search methods are very suitable for complex optimization problems. However, their search often results in a near-global optimal solution.

2.3 HYBRID APPROACHES FOR ED PROBLEM

In the past, nonconvex ED problems have been solved using many hybrid optimization methods. In Victoire and Jeyakumar (2004), the integration of the PSO and the sequential quadratic programming (SQP) techniques was proposed for solving the ED problem with valve-point effects. PSO is the main optimizer and the SQP was used to fine-tune the solution during each run. It was concluded that the proposed method outperforms and provides quality solutions as compared to other existing techniques for ED (considering valve-point effects).

In dos Santos Coelho and Mariani (2006), the proposed method combines the DE algorithm (with chaos sequence generation) and SQP techniques to deal with the ED problem with VPEs. The DE with chaos sequences acts as the global optimizer and the SQP is used to fine-tune the DE run in a sequential manner. The hybrid method and its variants were tested on two systems consisting of 13 and 40 thermal units. The proposed combined methods outperformed many other algorithms in solving ED problems with the valve-point effects.

In Wang, Chiou, and Liu (2007), the authors proposed an optimization algorithm called the self-tuning hybrid DE (self-tuning HDE). The HDE was used for solving ED problems which consider various generator constraints including ramp-rate limits, existence of POZs, and valve-point effects. The proposed method utilizes the concept of the 1/5 success rule of evolution strategies (ESs) in the original HDE to accelerate the search for finding the global optimum. Three test power systems (3-, 13-, and 40-unit) were employed to compare the performance of the proposed algorithm against those of the conventional GA, DE, and HDE algorithms. Numerical results indicated that the proposed self-tuning HDE algorithm outperforms the other three algorithms.

In He, Wang, and Mao (2008), an efficient method was proposed for solving the ED problem with incremental fuel cost function that takes into account the VPEs. The proposed method combines the GA, the SQP technique, uniform design technique, the maximum entropy principle, simplex crossover, and non-uniform mutation. The proposed hybrid technique uses GA as the main optimizer and the SQP to fine-tune the solution during each GA run. The results showed that the proposed hybrid GA improves the solution accuracy and reliability as compared to the other techniques implemented for the ED problem (which considers valve-point effects).

In Alsumait, Sykulski, and Al-Othman (2010), an approach based on a hybrid algorithm which consists of the GA, Pattern Search (PS), and SQP techniques was proposed for the ED problem. In that work, the GA was the main optimizer of the algorithm, whereas PS and SQP were used to fine-tune the results produced by the GA to increase the confidence of the solution. The proposed algorithm was applied to various test systems to assess its effectiveness. The outcome of that work was very encouraging, suggesting that the hybrid GA–PS–SQP algorithm is very efficient in solving the ED problem.

In Bhattacharya and Chattopadhyay (2010), a hybrid technique that combines the DE with biogeography-based optimization (DE-BBO) algorithm was proposed to solve both convex and nonconvex ED problems which considers transmission losses, ramp-rate limits, valve-point loading, and the existence of POZs. That work presented the combination of DE and BBO (DE-BBO) aiming to improve the solution quality and convergence speed. DE-BBO improved the search ability of the DE by utilizing the BBO algorithm effectively. Thus, the DE-BBO generates very promising candidate solutions. The effectiveness of the proposed algorithm was verified on four different test systems, both small and large scale. The author concluded that the proposed method is a very promising alternative for solving practical ED problems.

The aforementioned hybrid methods have become among the most effective search techniques for obtaining high-quality solutions. However, the hybrid methods may be slower and more algorithmically complicated than conventional methods since they combine several operations into one technique. Currently, nonconvex optimization problems are still a major challenge. Hence, there is a need for developing new techniques for solving these problems effectively.

2.4 MEAN-VARIANCE MAPPING OPTIMIZATION

Recently, mean-variance mapping optimization (MVMO) has been developed and introduced by Erlich, Venayagamoorthy, and Worawat (2010). This algorithm possesses conceptual similarities with other known metaheuristic algorithms which use three evolutionary operators: selection, mutation, and crossover. However, the special feature of MVMO is the mapping function applied for the mutation based on the mean and variance of n-best population saved in an archive. The original MVMO utilizes a single particle to start the search process. In order to enhance its global search ability, the search space of the MVMO is extended by initializing a set of particles which form a swarm variant of MVMO (MVMO^S) (Rueda & Erlich, 2013a). The subsequent improvement of this approach is the hybrid MVMO referred to as MVMO-SH (Rueda & Erlich, 2013b). The MVMO-SH adopts a swarm scheme of MVMO and incorporates LS components. Therefore, each particle has its own memory which is represented by the corresponding archive and mapping function. All particles are arranged according to their local best fitness and classified into two groups: good and bad particles. For each good particle, the parent assignment is done by considering the first ranked solution in its knowledge archive whereas a multiparent crossover is used to reorient each bad particle toward different subregions of the search space. An interior-point method (IPM) as an alternative to the SQP-based strategy is included for local improvement option. In this chapter, MVMO techniques are discussed with the aim of solving nonconvex ED problems. Thus far, MVMO has been successfully applied for solving different power system optimization problems (e.g., optimal reactive power dispatch problem [Nakawiro, Erlich, & Rueda, 2011], optimal of reactive sources in wind farm [Erlich, Nakawiro, & Martinez, 2011], optimal transmission expansion planning [Pringles & Rueda, 2012], identification of Gaussian mixture models [Gonzalez-Longatt, Rueda, Erlich, Bogdanov, & Villa, 2012], and identification of dynamic equivalents [Cepeda, Rueda, & Erlich, 2012]). The extension of MVMO called the swarm variant of MVMO ($MVMO^S$) has been successfully applied in some power system optimization problems. For instance, the optimal dispatch of reactive power sources and short-term transmission expansion planning.

2.5 ED OPTIMIZATION MODEL

The ED is described as an optimization problem which aims to minimize the total cost fuel function of thermal generating units in the power system (Wollenberg & Wood, 1996). The general formulation of the ED problem is as follows:

$$\begin{gathered}\text{Minimize} \rightarrow \text{Objective function} \\ \text{with respect to} \\ \text{Real power balance and generator capacity limits constraints}\end{gathered} \tag{2.1}$$

The objective function is written as follows:

$$F_{\mathrm{T}} = \sum_{i=1}^{N} F_i(P_i) \quad i = 1, 2, \ldots, N \tag{2.2}$$

where F_{T} is the total operational cost; P_i is the active power output of generator i; N is the total number of generating units in the system.

In the real power balance constraint as shown in Equation 2.3, the total active power output of the generating units must be equal to total power load demand P_{D} plus power loss P_{L}:

$$\sum_{i=1}^{N} P_i = P_{\mathrm{D}} + P_{\mathrm{L}} \tag{2.3}$$

where the power loss P_{L} is calculated by the below formula (Wollenberg & Wood, 1996):

$$P_{\mathrm{L}} = \sum_{i=1}^{N} \sum_{j=1}^{N} P_i B_{ij} P_j + \sum_{i=1}^{N} B_{0i} P_i + B_{00} \tag{2.4}$$

where B_{ij}, B_{0i}, and B_{00} are the B-matrix loss coefficients.

The active power output of each generating unit is restricted to the allowed limits, as shown in the generator capacity limits:

$$P_{i,\min} \leq P_i \leq P_{i,\max} \tag{2.5}$$

where $P_{i,\max}$ and $P_{i,\min}$ are the maximum and minimum power outputs of generator i, respectively.

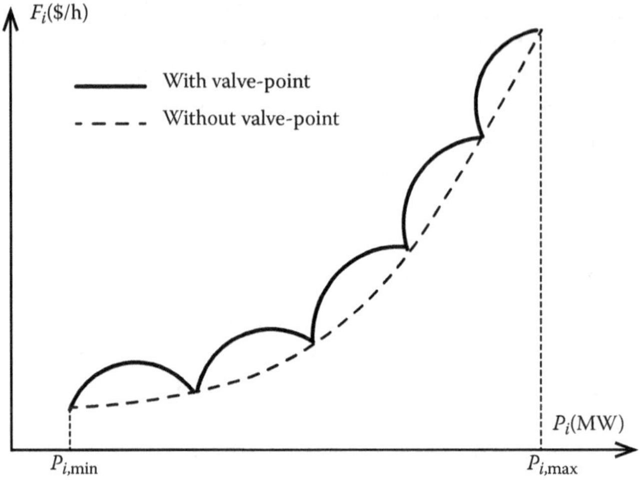

FIGURE 2.2 Cost function of generators with VPE.

Nonconvex ED is considered as a practical ED problem since it takes into account the VPEs, MF options, and the existence of POZs. The first problem is the ED with VPE. VPEs are the direct result of the practical operation of thermal generating units which produce ripple effects on the input–output curve as seen in Figure 2.2. This makes the fuel cost function highly nonlinear and nonsmooth containing multiple minima. The objective function of the generator is presented as follows (Al-Sumait et al., 2007):

$$F_i(P_i) = a_i + b_iP_i + c_iP_i^2 + \left|e_i \sin\left(f_i\left(P_{i,\min} - P_i\right)\right)\right| \tag{2.6}$$

where a_i, b_i, and c_i are the fuel cost coefficients of unit i; e_i and f_i are the fuel cost coefficients of unit i reflecting VPEs.

The second problem is ED with MF. Many thermal generating units in an electric power system are supplied with multiple-fuel sources such as coal, natural gas, and oil. This requires their fuel cost functions to be segmented as piecewise quadratic cost functions where each function reflects the effects of different fuel types. Figure 2.3 describes the fuel cost curve of the generator with MF. The ED problem with MF has piecewise quadratic cost functions which is a nonconvex optimization problem with multiple local optima (Chiang & Su, 2005):

$$F_i(P_i) = \begin{cases} a_{i1} + b_{i1}P_i + c_{i1}P_i^2, & \text{fuel 1, } P_{i,\min} \le P_i \le P_{i1} \\ a_{i2} + b_{i2}P_i + c_{i2}P_i^2, & \text{fuel 2, } P_{i1} < P_i \le P_{i2} \\ \cdots & \\ a_{ij} + b_{ij}P_i + c_{ij}P_i^2, & \text{fuel } j\text{, } P_{ij-1} < P_i \le P_{i,\max} \end{cases} \tag{2.7}$$

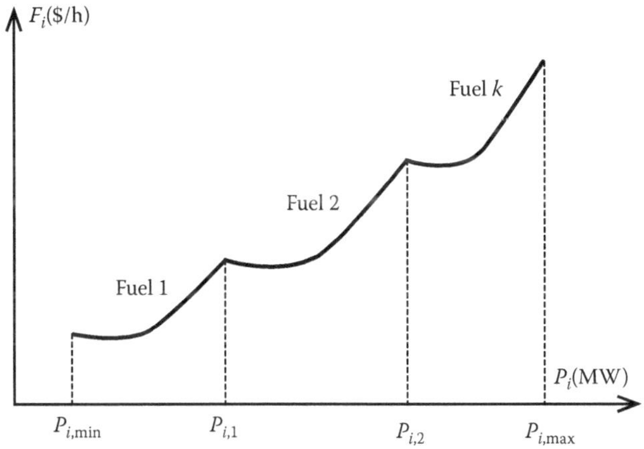

FIGURE 2.3 Cost function of generators with MF.

The objective of the ED problem with MF is to minimize the total cost (see Equation 2.2) where the fuel cost function for each generator is given in Equation 2.7 subject to the real power balance constraint (Equation 2.3) and generator capacity limits (see Equation 2.5).

The third problem is the ED with both VPE and MF. The solution of the practical ED problem is more accurate when the cost curve includes both VPE and MF. The objective function of each generating unit is given as follows (Dieu et al., 2011):

$$F_i(P_i)=\begin{cases} a_{i1}+b_{i1}P_i+c_{i1}P_i^2+\left|e_{i1}\cdot\sin\left(f_{i1}\cdot\left(P_{i1}^{\min}-P_{i1}\right)\right)\right|, & \text{for fuel 1, } P_i^{\min}\le P_i\le P_{i1} \\ a_{i2}+b_{i2}P_i+c_{i2}P_i^2+\left|e_{i2}\cdot\sin\left(f_{i2}\cdot\left(P_{i2}^{\min}-P_{i2}\right)\right)\right|, & \text{for fuel 2, } P_{i1}\le P_i\le P_{i2} \\ \vdots & \\ a_{ik}+b_{ik}P_i+c_{ik}P_i^2+\left|e_{ik}\cdot\sin\left(f_{ik}\cdot\left(P_{ik}^{\min}-P_{ik}\right)\right)\right|, & \text{for fuel } k,\ P_{ik-1}\le P_i\le P_i^{\max} \end{cases} \tag{2.8}$$

The objective of the ED problem with both VPE and MF is to minimize the total cost (Equation 2.1) where the fuel cost function for each generator is given in Equation 2.8 subject to the real power balance constraint (Equation 2.3) and generator capacity limits (Equation 2.5). The final problem is the ED with POZs. The POZs are the output power ranges at which the thermal generating units must avoid operations. This is because, at the POZs, the thermal generating units undergo undue vibrations of the turbine shaft and might bring damage to the shaft and bearings. The cost curve of the generator with POZs is divided into disjointed convex subregions as described in Figure 2.4.

The objective function of the ED problem with POZs can be formulated as a quadratic function. For units without POZs, only the equality constraint 2.3 and inequality constraint 2.5 are considered. As for units operating with POZs, more

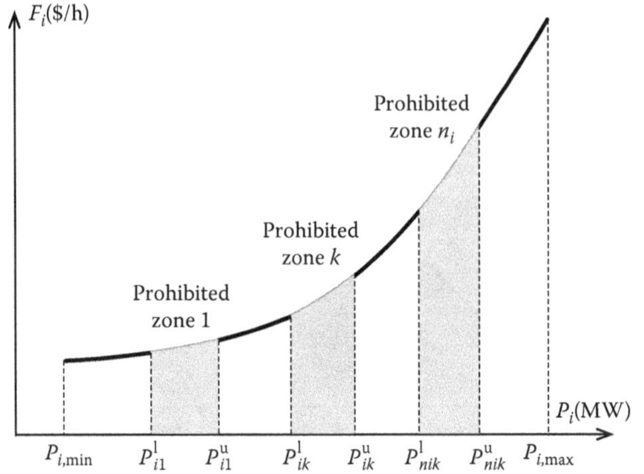

FIGURE 2.4 Cost function of generators with POZs.

constraints are added to the constraints mentioned above. The POZ constraints depict the feasible operating zones of each unit with POZs (Gaing, 2003):

$$P_i \in \begin{cases} P_{i,\min} \le P_i \le P^{l}_{i1} \\ P^{u}_{ik-1} \le P_i \le P^{l}_{ik}, \quad k = 2,\ldots,n_i \\ P^{u}_{in_i} \le P_i \le P_{i,\max} \end{cases} \tag{2.9}$$

where P^{l}_{ik} and P^{u}_{ik} are the lower and upper bounds for prohibited zone k of generator i, respectively. The ramp-rate limit constraints encapsulate the increased or decreased power output of a unit from its initial operating point to the next one. This limit states that the difference in power output should not exceed its ramp-up and ramp-down rate limits. The ramp-rate constraints are determined by Gaing (2003):

$$P_i - P_{i0} \le \text{UR}_i, \quad \text{if generation increases} \tag{2.10}$$

$$P_{i0} - P_i \le \text{DR}_i, \quad \text{if generation decreases} \tag{2.11}$$

where DR_i and UR_i are the ramp-down and ramp-up rate limits of unit i, respectively. P_{i0} is the initial output power of unit i.

2.6 METAHEURISTIC FRAMEWORK

2.6.1 Mean-Variance Mapping Optimization

MVMO is a new population-based metaheuristic optimization algorithm which was developed and introduced by István Erlich in 2010. The flowchart of MVMO is depicted in Figure 2.5 (Erlich et al., 2010).

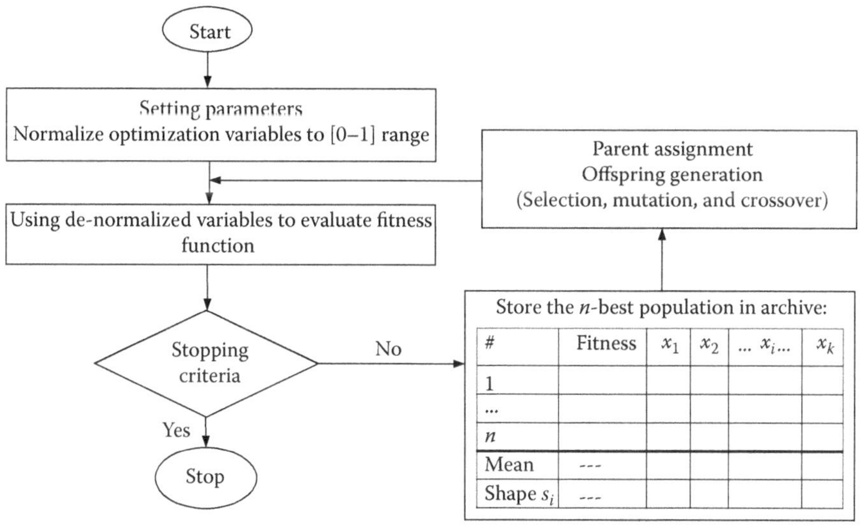

FIGURE 2.5 MVMO flowchart.

#	Fitness	x_1	x_2	... x_i ...	x_k
1					
...					
n					
Mean	---	$\bar{x}_1$	$\bar{x}_2$	$\bar{x}_i$	$\bar{x}_k$
Variance	---	v_1	v_2	v_i	v_k

FIGURE 2.6 The archive used to store n-best population.

The initial variables are normalized to the [0,1] bound as follows:

$$x\_normalized = rand(1,n\_var) \tag{2.12}$$

where n_var is the number of variable (generators); rand is the uniform random number between [0,1]. The search space of the algorithm is always restricted inside the [0,1] range. However, the function evaluation is carried out in the original scales $[x_{i,\min}, x_{i,\max}]$. The denormalization of the optimization variables is carried out as follows:

$$x_i = x_{i,\min} + (x_{i,\max} - x_{i,\min}) \cdot x\_normalized(i) \tag{2.13}$$

The MVMO algorithm possesses conceptual similarities to other known metaheuristic algorithms in three evolutionary operators (selection, mutation, and crossover operators). However, the special feature of MVMO is the mapping function applied for the mutation based on the mean and variance of the n-best population saved in an archive. The archive table is described in Figure 2.6. The archive size (n) is taken to be a minimum of two. The table of best individuals is filled up progressively over iterations in a descending order of the fitness. When the table is filled with n members, an update is performed only if the fitness of the new population is better than those in the table.

Mean $\overline{x}_i$ and variance v_i are computed from the archive as follows (Erlich et al., 2010):

$$\overline{x_i} = \frac{1}{n}\sum_{j=1}^{n} x_i(j) \tag{2.14}$$

$$v_i = \frac{1}{n}\sum_{j=1}^{n}\left(x_i(j) - \overline{x}_i\right)^2 \tag{2.15}$$

where j goes from 1 to n (archive size). At the beginning $\overline{x}_i$ corresponds with the initialized value of x_i and the variance v_i is set to 1. The individual with the best fitness, f_{best}, and its corresponding optimization values, x_{best}, are stored in the memory of the parent population for that iteration. This parent is used for creation of the next offspring. Three common evolutionary operations in offspring creation are: selection, mutation, and crossover operators. Among k variables of the optimization problem, m variables are selected for mutation. There are four strategies for the selection of variables as illustrated in Figure 2.7 (Erlich et al., 2010).

The variables are selected randomly by strategy 1. In stategy 2 and strategy 3, the variables are selected in the neighboring groups. Strategy 2 shows the group of variables which are moved forward in multiple steps. On the other hand, strategy 3 describes the group of variables which are moved forward in single steps. In strategy 4, the sequential selection is performed for the first variable. The rest of the variables are selected randomly. The m selected variables are then mutated via the mapping function as depicted in Figure 2.8.

Given a random number, $x_i^* \in [0,1]$, the transformation of x_i^* to x_i via the mapping function is calculated using Equation 2.16. The transformation mapping function, h, is calculated using the mean $\overline{x}$ and shape variables s_{i1} and s_{i2} as in Equation 2.18 (Erlich et al., 2010):

$$x_i = h_x + (1 - h_1 + h_0)\cdot x_i^* - h_0 \tag{2.16}$$

where h_x, h_1, and h_0 are the outputs of transformation mapping function based on different inputs as given by Erlich et al. (2010):

$$h_x = h(x = x_i^*),\ h_0 = h(x = 0),\ h_1 = h(x = 1) \tag{2.17}$$

$$h(\overline{x}_i, s_{i1}, s_{i2}, x) = \overline{x}_i \cdot (1 - \mathrm{e}^{-x\cdot s_{i1}}) + (1 - \overline{x}_i)\cdot \mathrm{e}^{-(1-x)\cdot s_{i2}} \tag{2.18}$$

where

$$s_i = -\ln(v_i)\cdot f_s \tag{2.19}$$

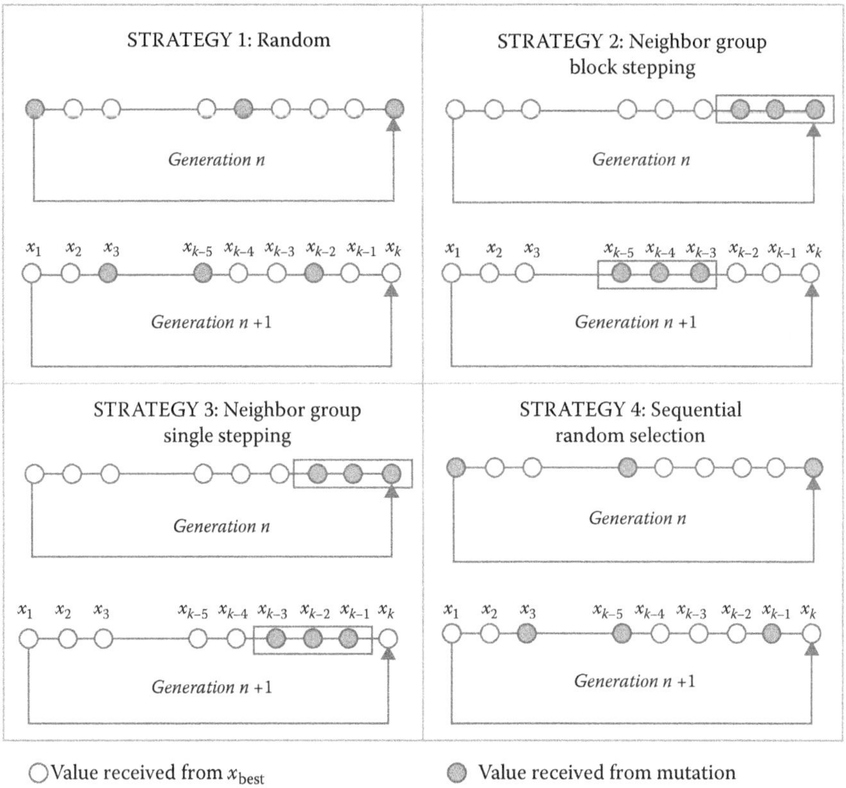

FIGURE 2.7 Strategies of variable selection ($m = 3$).

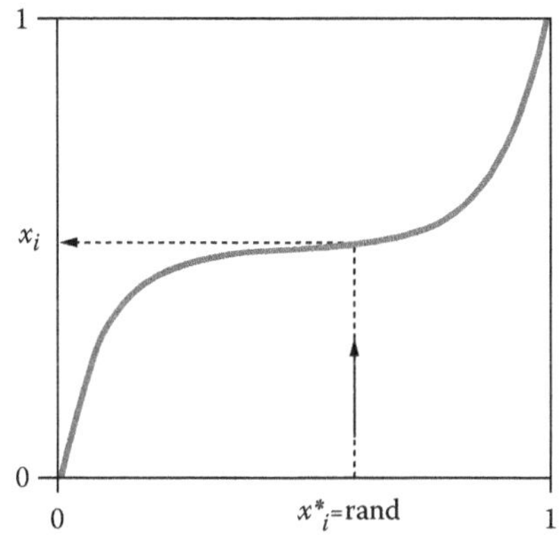

FIGURE 2.8 Variable mapping.

The scaling factor, f_s is a parameter which controls the shape of the mapping function during the search process. The shape variables s_{i1} and s_{i2} in Equation 2.18 are determined by using the following procedure (Erlich et al., 2010):

```
s_i1 = s_i2 = s_i
IF s_i > 0 THEN
  IF s_i > d_i
    d_i = d_i · Δd
  ELSE
    d_i = d_i / Δd
  END IF
IF rand() ≥ 0.5 THEN
    s_i1 = s_i ; s_i2 = d_i
    ELSE
    s_i1 = d_i ; s_i2 = s_i
    END IF
END IF
```

The initial values of d_i are set for all variables prior to the search operation. The values of d_i are usually defined in the range of 1 to 5 for good initial performance. Zero variance can occur when all variables of x_i in the archive are identical. In this case, the previous nonzero value is used further. Sometimes the variance oscillates over a wide range. By using the factor d_i instead of s_i (which is a function of variance), a smoothing effect is achieved.

The crossover operation is done for the remaining unmutated dimensions where the genes of the parents are inherited. In other words, the values of these unmutated dimensions are clones of the parent. Here, the crossover is done by direct cloning of certain genes. In this way, the offsprings are created by combining the vector x_{best} and vector of m mutated dimensions.

2.6.2 Swarm-Based Mean-Variance Mapping Optimization

MVMO^S is an extension of the original version MVMO. The difference between MVMO and MVMO^S is the initial search process with particles. MVMO starts the search with a single particle while MVMO^S starts the search with a set of particles. MVMO is extended to two parameters: the scaling factor f_s and variable increment Δd parameter to enhance its mapping. Therefore, the global search ability of the MVMO^S is strengthened. In MVMO^S, the factor f_s in Equation 2.19 is extended due to the need for exploring the search space more globally at the beginning of the iterations. At the end of the iterations, the focus should be on exploitation procedures. In Rueda and Erlich (2013a), the factor f_s is given as follows:

$$f_s = f_s^* \cdot (1 + \text{rand}()) \tag{2.20}$$

where

$$f_s^* = f_{s\_ini}^* + \left(\frac{i}{i_{final}}\right)^2 \left(f_{s\_final}^* - f_{s\_ini}^*\right) \quad (2.21)$$

In Equation 2.21, f_s^* denotes the smallest value of f_s and the variable i represents the iteration number. $f_{s\_ini}^*$ and $f_{s\_final}^*$ are the initial and final values of f_s^*, respectively. The recommended range of $f_{s\_ini}^*$ is from 0.9 to 1.0, and range of $f_{s\_final}^*$ is from 1.0 to 3.0. When $f_{s\_final}^* = f_{s\_ini}^* = 1$, it means that the option for controlling the f_s factor is not used. In MVMO^S algorithm, the shape variables s_{i1} and s_{i2} in Equation 2.18 are determined by using the following procedure (Rueda & Erlich, 2013c):

```
s_i1 = s_i2 = s_i
IF s_i > 0 THEN
Δd = (1 + Δd_0) + 2 · Δd_0 (rand() – 0.5)
  IF s_i > d_i
    d_i = d_i · Δd
  ELSE
    d_i = d_i /Δd
  END IF
  IF rand() ≥ 0.5 THEN
    s_i1 = s_i ; s_i2 = d_i
    ELSE
    s_i1 = d_i ; s_i2 = s_i
    END IF
END IF
```

The above procedure fully exploits the asymmetric characteristic of the mapping function by using different values for s_{i1} and s_{i2}. Δd_0 (calculated in Equation 2.22) is allowed to decrease from 0.4 to 0.01 (Rueda & Erlich, 2013c).

$$\Delta d_0 = \Delta d_0^{ini} + \left(\frac{i}{i_{final}}\right)^2 \left(\Delta d_0^{final} - \Delta d_0^{ini}\right) \quad (2.22)$$

where d_i is the initial smoothing factor. Δd_0^{ini} and Δd_0^{final} are the initial and final smoothing factor increments, respectively.

MVMO^S utilizes swarm implementation to enhance its power for global search. This is done by starting the search process of the classical MVMO with a set of N_p particles, each having its own memory and represented by the corresponding archive and mapping function. At the beginning of the optimization process, each particle performs m steps independently to collect a set of reliable individual solutions. Then, the particles start communicating and exchanging information.

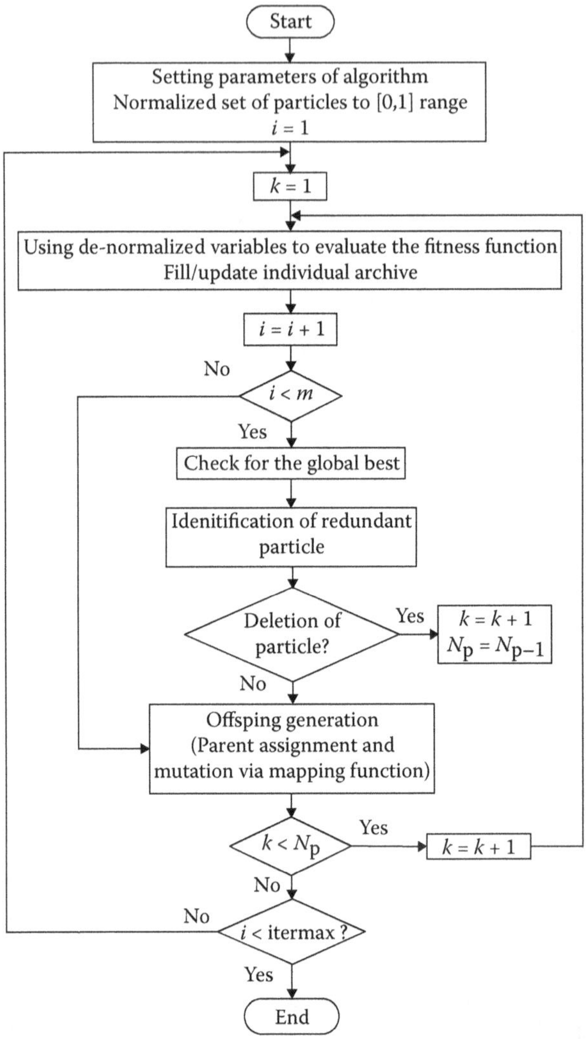

FIGURE 2.9 MVMO^S flowchart.

The flowchart of MVMO^S is depicted in Figure 2.9, where i and k are the function evaluation and particle counters, respectively. The parameters m and N_p are for maximum number of independent runs and total number of particles, respectively.

2.6.3 Hybrid Variant MVMO

The hydrid variant mean-variance mapping optimization (MVMO-SH) is a subsequent improvement of the original MVMO-SH. It adopts a swarm scheme of MVMO and incorporates LS and multiparent crossover strategies to increase the innate power of the global search ability of MVMO. The flowchart of MVMO-SH is shown in Figure 2.10.

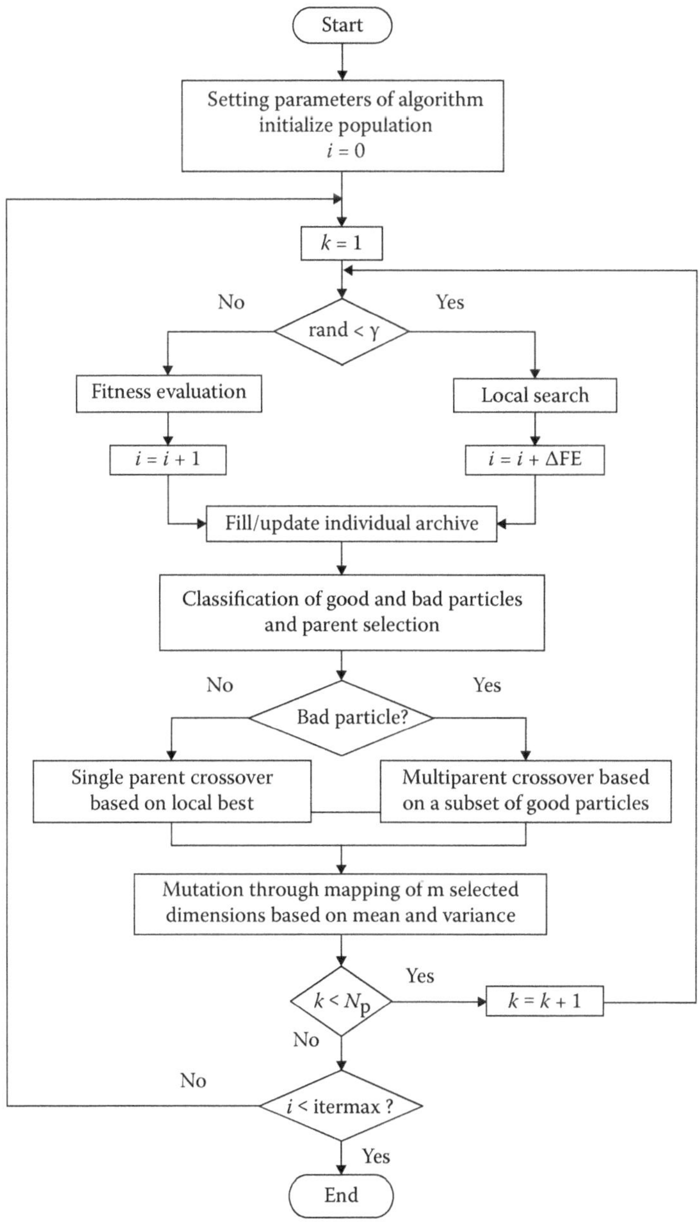

FIGURE 2.10 MVMO-SH flowchart.

As seen in Figure 2.10, the fitness evaluation and particle counters are denoted by i and k, whereas N_p and ΔFE represent the total number of particles and the number of fitness evaluations required by LS, respectively. The core of MVMO-SH encompasses two nested loops which are carried out after the initialization stage. The outer loop is finished once the maximum number of iterations is reached. The procedure

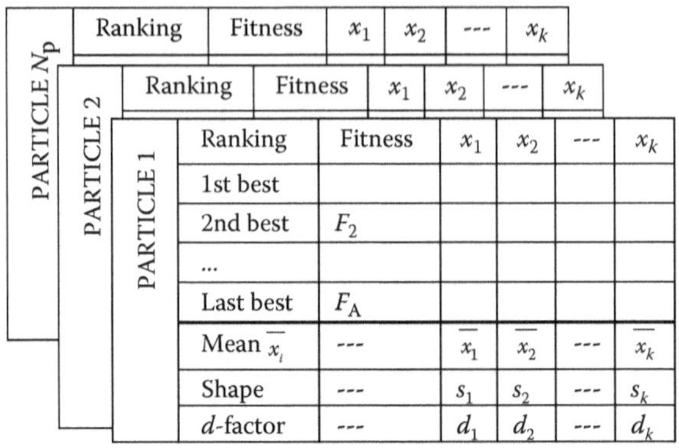

FIGURE 2.11 Structure of the solution archive.

of the inner loop comprises five steps: fitness evaluation/LS, fill/update the archive, classification of particles, parent selection, and offspring generation. The variables of the optimization problem are recalculated from [0,1] to their original boundaries before fitness evaluation or LS is carried out. An IPM or a SQP-based strategy is included as a local improvement option. IMP or SQP is performed randomly for any child of the population.

The solution archive stores the n-best achieved offspring of each particle in a descending order of fitness. For each particle, its archive is only updated once the new better solution is produced after every step of fitness evaluation or LS. The archive is illustrated in Figure 2.11.

At least two fitness functions are independently evaluated by each particle in the early stage of the search process, and those solutions are saved in the archive. All particles are ranked based on their local best fitness. Among them, a set of good particles G_p is determined as in Equation 2.23, and the remainder is classified as a set of bad particles (N_p–G_p) (Rueda & Erlich, 2013c).

$$G_p = \text{round}(N_p \cdot g_p) \tag{2.23}$$

where

$$g_p = g_{p\_ini} + \frac{i}{i_{final}}(g_{p\_final} - g_{p\_ini}) \tag{2.24}$$

where g_{p_ini} and g_{p_final} are the maximum and minimum percentage of good particles, respectively.

For each good particle, the first ranked solution in its particular knowledge archive is chosen as the parent for the next offspring. For each bad particle, a multiparent strategy is used to produce the parent as in Equation 2.25. The details of parent selection are described in Rueda and Erlich (2013c).

$$x_k^{parent} = x_{RG}^{best} + \beta\left(x_{GB}^{best} - x_{LG}^{best}\right) \tag{2.25}$$

The x_{RG}^{best} is randomly selected between the x_{GB}^{best} and x_{LG}^{best}, which represent the first and the last global best in the group of good particles, respectively. The β factor is a random number which is calculated as in Equation 2.26. This number is recalculated if any element of the vector x_k^{parent} is outside the [0,1] bound (Rueda & Erlich, 2013c).

$$\beta = 2\left(\text{rand} + 0.5\left(\frac{i}{i_{\text{final}}}\right)^2 - 0.5\right) \tag{2.26}$$

In the MVMO-SH variable selection, the m number of dimensions to be selected for mutation is progressively decreased as follows (Rueda & Erlich, 2013c):

$$m = \text{round}(m_{\text{final}} + \text{rand}(m^* - m_{\text{final}})) \tag{2.27}$$

where

$$m^* = \text{round}(m_{\text{ini}} - \left(\frac{i}{i_{\text{final}}}\right)^2 (m_{\text{ini}} - m_{\text{final}})) \tag{2.28}$$

where m_{ini} and m_{final} are the initial and final number of variables selected for mutation, respectively.

2.7 CONCLUSION

This chapter provides an overview of the optimization methods and techniques employed for solving the nonconvex ED problem which takes into account VPE, MF, and POZs with ramp-rate limit as well as transmission network losses. These methods are classified into classical calculus-based methods, artificial intelligence techniques, and hybrid methods. The classical methods are: lambda iteration, Newton's method, gradient search, dynamic programming, linear programming, nonlinear programming, and quadratic programming. These methods have proven their effectiveness in solving the ED problem with smooth and convex cost functions. However, most of these methods fail to obtain a global optimal solution for the ED problem with nonsmooth or nonconvex cost functions. Recently, artificial intelligence techniques (such as HNN, EP, DE, GA, ACO, BFO, ABC algorithm, and PSO) have been successfully applied for solving nonsmooth or nonconvex ED problems. This is due to their capacity to efficiently find the global optimal solution. However, artificial intelligence techniques often take too much computational time to reach the near-global optimum. Hybrid methods which combine two or more optimization techniques are found to be more effective in finding the global optimal solution for ED problems with nonsmooth or nonconvex cost functions. However, the hybrid methods may be slower and more complicated than conventional methods due to the increase in algorithmic complexity.

This chapter has also presented an overview of the proposed MVMO techniques for solving nonconvex ED problems. The MVMO algorithm is a population-based stochastic optimization technique similar to ACO, BFO, PSO, and ABC. These mentioned swarm intelligence-based techniques have thus far been successfully applied for solving the nonconvex ED problems. So far, MVMO techniques have been implemented effectively for solving different power system optimization problems such as optimal reactive power dispatch problem, optimal reactive sources in wind farm, optimal transmission expansion planning, identification of Gaussian mixture models, identification of dynamic equivalents, and short-term transmission expansion planning. The MVMO techniques such as $MVMO^S$ and MVMO-SH have never been applied for nonconvex ED problems. Therefore, these methods have a great potential for effectively solving the nonconvex ED problem.

REFERENCES

Abedinia, O., Amjady, N., Ghasemi, A., & Hejrati, Z. (2013). Solution of economic load dispatch problem via hybrid particle swarm optimization with time-varying acceleration coefficients and bacteria foraging algorithm techniques. *International Transactions on Electrical Energy Systems*, *23*(8), 1504–1522.

Al-Sumait, J., Al-Othman, A., & Sykulski, J. (2007). Application of pattern search method to power system valve-point economic load dispatch. *International Journal of Electrical Power & Energy Systems*, *29*(10), 720–730.

Alsumait, J., Sykulski, J., & Al-Othman, A. (2010). A hybrid GA–PS–SQP method to solve power system valve-point economic dispatch problems. *Applied Energy*, *87*(5), 1773–1781.

Aydın, D., & Özyön, S. (2013). Solution to nonconvex economic dispatch problem with valve point effects by incremental artificial bee colony with local search. *Applied Soft Computing*, *13*(5), 2456–2466.

Bhattacharya, A., & Chattopadhyay, P. K. (2010). Hybrid differential evolution with biogeography-based optimization for solution of economic load dispatch. *Power Systems, IEEE Transactions*, *25*(4), 1955–1964.

Cepeda, J. C., Rueda, J. L., & Erlich, I. (2012). Identification of dynamic equivalents based on heuristic optimization for smart grid applications. *Evolutionary Computation (CEC), 2012 IEEE Congress* (pp. 1–8), Piscataway, NJ: IEEE.

Chaturvedi, K. T., Pandit, M., & Srivastava, L. (2008). Self-organizing hierarchical particle swarm optimization for nonconvex economic dispatch. *Power Systems, IEEE Transactions*, *23*(3), 1079–1087.

Chiang, C.-L. (2005). Improved genetic algorithm for power economic dispatch of units with valve-point effects and multiple fuels. *Power Systems, IEEE Transactions*, *20*(4), 1690–1699.

Chiang, C.-L., & Su, C.-T. (2005). Adaptive-improved genetic algorithm for the economic dispatch of units with multiple fuel options. *Cybernetics and Systems: An Internationl Journal*, *36*(7), 687–704.

Dieu, V. N., Schegner, P., & Ongsakul, W. (2011). A newly improved particle swarm optimization for economic dispatch with valve point loading effects. *Power and Energy Society General Meeting, 2011 IEEE* (pp. 1–8), Piscataway, NJ: IEEE.

Dieu, V. N., Schegner, P., & Ongsakul, W. (2013). Pseudo-gradient based particle swarm optimization method for nonconvex economic dispatch. *Power, control and optimization* (pp. 1–27). Switzerland: Springer International Publishing.

dos Santos Coelho, L., & Mariani, V. C. (2006). Combining of chaotic differential evolution and quadratic programming for economic dispatch optimization with valve-point effect. *Power Systems, IEEE Transactions*, *21*(2), 989–996.

Erlich, I., Nakawiro, W., & Martinez, M. (2011). Optimal dispatch of reactive sources in wind farms. *Power and Energy Society General Meeting, 2011 IEEE* (pp. 1–7), Piscataway, NJ: IEEE.

Erlich, I., Venayagamoorthy, G. K., & Worawat, N. (2010). A mean-variance optimization algorithm. *IEEE Congress on Evolutionary Computation* (pp. 1–6), Piscataway, NJ: IEEE.

Fan, J.-Y., & Zhang, L. (1998). Real-time economic dispatch with line flow and emission constraints using quadratic programming. *Power Systems, IEEE Transactions*, *13*(2), 320–325.

Gaing, Z.-L. (2003). Particle swarm optimization to solving the economic dispatch considering the generator constraints. *Power Systems, IEEE Transactions*, *18*(3), 1187–1195.

Gonzalez-Longatt, F., Rueda, J., Erlich, I., Bogdanov, D., & Villa, W. (2012). Identification of gaussian mixture model using mean variance mapping optimization: Venezuelan case. *Innovative Smart Grid Technologies (ISGT Europe), 2012 3rd IEEE PES International Conference and Exhibition* (pp. 1–6), Piscataway, NJ: IEEE.

He, D.-K., Wang, F.-L., & Mao, Z.-Z. (2008). Hybrid genetic algorithm for economic dispatch with valve-point effect. *Electric Power Systems Research*, *78*(4), 626–633.

Hemamalini, S., & Simon, S. (2009). Maclaurin series-based Lagrangian method for economic dispatch with valve-point effect. *Generation, Transmission & Distribution, IET*, *3*(9), 859–871.

Hemamalini, S., & Simon, S. P. (2010). Artificial bee colony algorithm for economic load dispatch problem with nonsmooth cost functions. *Electric Power Components and Systems*, *38*(7), 786–803.

Holland, J. H. (1975). *Adaptation in natural and artificial systems: An introductory analysis with applications to biology, control, and artificial intelligence*. Cambridge, MA: University of Michigan Press.

Karaboga, D. (2005). An idea based on honey bee swarm for numerical optimization: Technical report-TR06. Erciyes University, Engineering Faculty, Computer Engineering Department.

Kuo, C.-C. (2008). A novel coding scheme for practical economic dispatch by modified particle swarm approach. *Power Systems, IEEE Transactions*, *23*(4), 1825–1835.

Lee, K. Y., Sode-Yome, A., & Park, J. H. (1998). Adaptive Hopfield neural networks for economic load dispatch. *Power Systems, IEEE Transactions*, *13*(2), 519–526.

Mahor, A., Prasad, V., & Rangnekar, S. (2009). Economic dispatch using particle swarm optimization: A review. *Renewable and Sustainable Energy Reviews*, *13*(8), 2134–2141.

Nakawiro, W., Erlich, I., & Rueda, J. L. (2011). A novel optimization algorithm for optimal reactive power dispatch: A comparative study. *Electric Utility Deregulation and Restructuring and Power Technologies (DRPT), 2011 4th International Conference* (pp. 1555–1561), Piscataway, NJ: IEEE.

Nanda, J., Hari, L., & Kothari, M. (1994). Economic emission load dispatch with line flow constraints using a classical technique. *IEE Proceedings-Generation, Transmission and Distribution*, *141*(1), 1–10.

Noman, N., & Iba, H. (2008). Differential evolution for economic load dispatch problems. *Electric Power Systems Research*, *78*(8), 1322–1331.

Padmanabhan, B., Sivakumar, R., Jasper, J., & Victoire, T. A. A. (2011). Bacterial foraging approach to economic load dispatch problem with nonconvex cost function. *Swarm, evolutionary, and memetic computing* (pp. 577–584). Berlin, Heidelberg: Springer-Verlag.

Parikh, J., & Chattopadhyay, D. (1996). A multi-area linear programming approach for analysis of economic operation of the Indian power system. *Power Systems, IEEE Transactions*, *11*(1), 52–58.

Park, J., Kim, Y., Eom, I., & Lee, K. (1993). Economic load dispatch for piecewise quadratic cost function using Hopfield neural network. *Power Systems, IEEE Transactions*, *8*(3), 1030–1038.

Park, J.-B., Lee, K.-S., Shin, J.-R., & Lee, K. Y. (2005). A particle swarm optimization for economic dispatch with nonsmooth cost functions. *Power Systems, IEEE Transactions*, *20*(1), 34–42.

Pothiya, S., Ngamroo, I., & Kongprawechnon, W. (2010). Ant colony optimisation for economic dispatch problem with nonsmooth cost functions. *International Journal of Electrical Power & Energy Systems*, *32*(5), 478–487.

Pringles, R. M., & Rueda, J. L. (2012). Optimal transmission expansion planning using mean-variance mapping optimization. *Transmission and Distribution: Latin America Conference and Exposition (T&D-LA), 2012 Sixth IEEE/PES* (pp. 1–8), Piscataway, NJ: IEEE.

Rahimullah, B. N. S., Ramlan, E. I., & Rahman, T. K. A. (2003, December 15–16, 2003). Evolutionary approach for solving economic dispatch in power system. *Proceedings of National Power Engineering Conference, 2003. PECon 2003.* (pp. 32–36), Piscataway, NJ: IEEE.

Rueda, J. L., & Erlich, I. (2013a). Evaluation of the mean-variance mapping optimization for solving multimodal problems. *Swarm Intelligence (SIS), 2013 IEEE Symposium* (pp. 7–14), Piscataway, NJ: IEEE.

Rueda, J. L., & Erlich, I. (2013b). Hybrid mean-variance mapping optimization for solving the IEEE-CEC 2013 competition problems. *Evolutionary Computation (CEC), 2013 IEEE Congress* (pp. 1664–1671), Piscataway, NJ: IEEE.

Rueda, J. L., & Erlich, I. (2013c). Optimal dispatch of reactive power sources by using MVMOs optimization. *Computational Intelligence Applications in Smart Grid (CIASG), 2013 IEEE Symposium* (pp. 29–36), Piscataway, NJ: IEEE.

Selvakumar, A. I., & Thanushkodi, K. (2007). A new particle swarm optimization solution to nonconvex economic dispatch problems. *Power Systems, IEEE Transactions*, *22*(1), 42–51.

Sinha, N., Chakrabarti, R., & Chattopadhyay, P. (2003). Evolutionary programming techniques for economic load dispatch. *Evolutionary Computation, IEEE Transactions*, *7*(1), 83–94.

Storn, R., & Price, K. (1997). Differential evolution—A simple and efficient heuristic for global optimization over continuous spaces. *Journal of Global Optimization*, *11*(4), 341–359.

Victoire, T. A. A., & Jeyakumar, A. E. (2004). Hybrid PSO–SQP for economic dispatch with valve-point effect. *Electric Power Systems Research*, *71*(1), 51–59.

Wang, S.-K., Chiou, J.-P., & Liu, C. (2007). Nonsmooth/nonconvex economic dispatch by a novel hybrid differential evolution algorithm. *Generation, Transmission & Distribution, IET*, *1*(5), 793–803.

Wollenberg, B., & Wood, A. (1996). *Power generation, operation and control.* New York: John Wiley & Sons.

3 Smart Charging Optimization of Plug-in Hybrid Electric Vehicles

Plug-in hybrid electric vehicle (PHEV) optimization is an emerging field which is increasingly being implemented in many charging infrastructures on a global scale. A near-comprehensive understanding of smart charging is crucial for the large participation of PHEVs. Proper charging can reassure PHEV users, lessen their "range anxiety," and switch into the new green vehicle revolution with less carbon-dioxide emissions. In this chapter, the authors present swarm intelligence-based optimizations to optimize the smart charging of PHEVs in charging infrastructures.

3.1 INTRODUCTION

Recent studies on renewable technologies for the transportation sector are gaining much attention among various research groups. As a result, PHEVs have a bright future because of their charge storage system and smart charging facilities from customary grid system. Several researchers have proven that significant reductions in greenhouse gas emissions and reduced dependence on oil could be accomplished by electrification of the transport sector (Caramanis & Foster, 2009). Future transportation sector will depend much on the advancement of this emerging field of vehicle optimization. Indeed, the adoption of hybrid electric vehicles (HEVs) has brought significant market success over the past decade. Vehicles can be classified into three groups: internal combustion engine vehicles (ICEVs), HEVs, and all-electric vehicles (AEVs) (Tie & Tan, 2013). PHEVs which have been very recently introduced promise to boost up the overall fuel efficiency by holding a higher-capacity battery system, which can be directly charged from conventional power grid system that helps the vehicles to operate continuously in "all-electric-range" (AER). AEVs are a kind of transport that uses electric power as the only source to run the system. PHEVs with a connection to the smart grid can own all of these strategies. Hence, the widely extended adoption of PHEVs might play a significant role in the alternative energy integration into traditional grid systems (Lund & Kempton, 2008). There is a need for efficient mechanisms and algorithms for smart grid technologies in order to solve highly diverse problems such as energy management, cost reduction, efficient charging station, and so on, with different objectives and system constraints (Hota, Juvvanapudi, & Bajpai, 2014).

According to statistics of the Electric Power Research Institute (EPRI), about 62% of the entire US vehicles will comprise PHEVs within the year 2050 (Soares et al., 2013). Moreover, there is an increasing demand to implement this technology on the electric grid system. Large numbers of PHEVs have the capability to make threats to the stability of the power system. For example, in order to avoid disturbance when several thousand PHEVs are introduced into the system over a small period, the load on the power grid will need to be managed very carefully. One of the main targets is to facilitate proper communication between the power grid and the PHEV. For the maximization of customer contentment and minimization of burdens on the grid, a complicated control appliance will need to be addressed in order to govern multiple battery loads from a number of PHEVs properly (Su & Chow, 2012a). The total demand pattern will also have an important impact on the electricity production due to differences in the needs of the PHEVs parked in the deck at a certain time (Su & Chow, 2011). Proper management can ensure strain minimization of the grid and enhance the transmission and generation of electric power supply. The control of PHEV charging depending on the locations can be classified into two groups: household charging and public charging. The proposed optimization focuses on the public charging station for plug-in vehicles because most of the PHEV charging is expected to take place in public charging locations (Su & Chow, 2012a). Wide penetration of PHEVs in the market depends on a well-organized charging infrastructure. The power demand from this new load will put extra stress on the traditional power grid (Morrow, Karner, & Francfort, 2008). As a result, a good number of PHEV charging stations with suitable facilities are essential to be built for recharging electric vehicles (EVs); for this some strategies have been proposed by the researchers (Mayfield, 2012). Charging stations are needed to be built at workplaces, markets/shopping malls, and home. Boyle (2007) proposed the necessity of building new smart charging stations with effective communication among utilities along with substation control infrastructure in view of grid stability and proper energy utilization. Furthermore, ample energy storage, cost minimization, quality of services (QoS), and intelligent charging stations for optimal power are underway (Hess et al., 2012). In this wake, numerous techniques and methods were proposed for deployment of PHEV charging stations (Li, Sahinoglu, Tao, & Teo, 2010).

One of the main targets is to facilitate proper interaction between the power grid and the PHEV. For the maximization of customer satisfaction and minimization of burdens on the grid, a complicated control mechanism will need to be addressed in order to govern multiple battery loads from a number of PHEVs appropriately (Su & Chow, 2012b). Charging infrastructures are essential in order to facilitate the large-scale penetration of PHEVs. Different computational intelligence-based methods have been used by some researchers for charging station optimization of PHEVs. Most of them applied traditional methods which are needed to be improved further.

Swarm intelligence came from the mimicking of living colonies such as those of ants, birds, and fish in nature, which show unparalleled excellence in swarm behavior rather than being single in food seeking or nest building. Drawing inspiration from this, researchers designed many algorithms simulating colony living, such as ant colony algorithm, particle swarm optimization (PSO) algorithm, artificial bee colony (ABC) algorithm, and gravitational search algorithm, which show excellent performance in dealing with complex optimization problems (Jia-Zhao, Yu-Xiang, & Yin-Sheng, 2012). The intrinsic characteristics of all the population-based metaheuristic algorithms such as PSO and accelerated particle swarm optimization (APSO) are to maintain a good compromise between exploration and exploitation in order to solve the complex optimization problems (Rahman, Vasant, Singh, & Abdullah-Al-Wadud, 2014b).

PSO is based on two fundamental disciplines: social science and computer science. In addition, PSO uses the swarm intelligence concept, which is the property of a system, whereby the collective behaviors of unsophisticated agents that are interacting locally with their environment create coherent global functional patterns. PSO algorithm has been successfully used for solving many problems related to power systems (del Valle, Venayagamoorthy, Mohagheghi, Hernandez, & Harley, 2008) such as voltage security, optimal power flow, power system operation and planning, dynamic security, power quality, unit commitment, reactive power control, capacitor placement, and optimizing controller parameters.

Moreover, APSO-based optimizations have already been studied by the researchers for optimal design of substation grounding grid, nonconvex optimization (Ganesan, Vasant, & Elamvazuthy, 2012), performance analysis of radar waveform, design of frame structures, dual channel speech enhancement (Prajna, Rao, Reddy, & Maheswari, 2014), synthesis gas production, and a faster path planner (Mohamed, Lee, Hsu, & Nath, 2012).

The performance of a PHEV depends on proper consumption of electric power which is solely affected by the battery state-of-charge (SoC). In PHEVs, a key parameter is the SoC of the battery as it is a measure of the amount of electrical energy stored in it. It is analogous to a fuel gauge on a conventional internal combustion (IC) car (Chiasson & Vairamohan, 2005). SoC determination becomes a progressively dynamic issue in all the areas that include a battery. Previous operation policies made use of voltage limits only to guard the battery against deep discharge and overcharge. Currently, battery operation is changing to what could rather be called battery management than simply protection. For this improved battery control, the battery SoC is a key factor indeed (Piller, Perrin, & Jossen, 2001).

A charging station is one way that the operator of an electrical power grid can adapt energy production to energy consumption, both of which can vary randomly over time. Basically, PHEVs in a charging station are charged during times when production exceeds consumption and are discharged at times when consumption exceeds production. There is a need of in-depth study on maximization of average SoC in order to facilitate intelligent energy allocation for PHEVs in a charging station.

The purpose of this chapter is to optimize SoC, with respect to total cost, charging time, and present SoC. Two swarm intelligence-based methods, PSO and APSO, were applied for solving the optimization problem.

3.2 DEVELOPMENT OF FITNESS FUNCTION

The idea behind smart charging is to charge the vehicle when it is most favorable, which could be when electricity price and demand are lowest and when there is excess capacity (Su & Chow, 2012b).

Suppose, there is a charging station with the capacity of total power P. Total N numbers of PHEVs need to be served in a day (24 h). The proposed system should allow PHEVs to leave the charging station before their expected leaving time for making the system more effective. It is worth mentioning that each PHEV is regarded to be plugged-in to the charging station once. The main aim is to allocate power intelligently for each PHEV coming to the charging station. The SoC is the main parameter which needs to be maximized in order to allocate power efficiently. For this, the fitness function considered in this chapter is the maximization of average SoC and thus allocate energy for PHEVs at the next time step. The constraints considered are: charging time, present SoC, and price of the energy (Rahman, Vasant, Singh, & Abdullah-Al-Wadud, 2014a).

The fitness function is defined as

$$\text{Max } J(k) = \sum_i w_i(k)\text{SoC}_i(k+1) \tag{3.1}$$

$$w_i(k) = f\left(C_{\text{r},i}(k), T_{\text{r},i}(k), D_i(k)\right) \tag{3.2}$$

$$C_{\text{r},i}(k) = \left(1 - \text{SoC}_i(k)\right) \cdot C_i \tag{3.3}$$

where $C_{\text{r},i}(k)$ is the battery capacity (remaining) needed to be filled for i number of PHEVs at time step k; C_i is the battery capacity (rated) of the i number of PHEVs; remaining time for charging a particular PHEV at time step k is articulated as $T_{\text{r},i}(k)$; the price difference between the real-time energy price and the price that a specific customer at the i number of PHEV charger is willing to pay at time step k is presented by $D_i(k)$; $w_i(k)$ is the charging weighting term of the i number of PHEVs at time step k (a function of charging time, present SoC, and price of the energy); $\text{SoC}_i(k + 1)$ is the SoC of the i number of PHEVs at time step $k + 1$.

Here, the weighting term indicates a bonus proportional to the attributes of a specific PHEV. For example, if a PHEV has a lower initial SoC and less charging time (remaining), but the driver is willing to pay a higher price, the system will provide more power to this particular PHEV battery charger:

$$w_i(k)\alpha\left[C_{\text{r},i}(k) + D_i(k) + 1/T_{\text{r},i}(k)\right] \tag{3.4}$$

The charging current is also assumed to be constant over Δt.

$$\left[\text{SoC}_i(k+1) - \text{SoC}_i(k)\right] \cdot \text{Cap}_i = Q_i = I_i(k)\Delta t \tag{3.5}$$

$$\text{SoC}_i(k+1) = \text{SoC}_i(k) + I_i(k)\Delta t/\text{Cap}_i \tag{3.6}$$

where the sample time Δt is defined by the charging station operators, and $I_i(k)$ is the charging current over Δt.

The battery model is regarded as a capacitor circuit, where C_i is the capacitance of battery (Farad). The model is defined as

$$C_i \cdot \frac{dV_i}{dt} = I_i \tag{3.7}$$

Therefore, over a small time interval, one can assume the change of voltage to be linear,

$$C_i \cdot \left[V_i(k+1) - V_i(k)\right] / \Delta t = I_i \tag{3.8}$$

$$V_i(k+1) - V_i(k) = I_i \Delta t / C_i \tag{3.9}$$

As the decision variable used here is the allocated power to the PHEVs, by replacing $I_i(k)$ with $P_i(k)$ the objective function finally becomes:

$$\mathrm{SoC}_i(k+1) = \mathrm{SoC}_i(k) + \frac{P_i(k)\Delta t}{0.5 \cdot C_i \cdot \left[\sqrt{\frac{2P_i(k)\Delta t}{C_i} + V_i^2(k)} + V_i(k)\right]} \tag{3.10}$$

Power obtained from the utility (P_{utility}) and the maximum power ($P_{i,\max}$) absorbed by a specific PHEV are the primary energy constraints being considered in this chapter. The overall charging efficiency of a particular charging station is described by η. From the system point of view, charging efficiency is supposed to be constant at any given time step. Maximum battery SoC limit for the i number of PHEVs is $\mathrm{SoC}_{i,\max}$. When SoC_i reaches the values close to $\mathrm{SoC}_{i,\max}$, the i number of battery charger shifts to a standby mode. The SoC ramp rate is confined within limits by the constraint $\Delta\mathrm{SoC}_{\max}$. The overall control system has changed the state when (i) system utility data updates; (ii) a new PHEV is plugged in; and (iii) time period Δt has periodically passed.

Table 3.1 shows all the objective function parameters that were tuned for performing the optimization. There are totally three kinds of parameters: fixed, variables, and constraints. Total charging time is fixed to 20 min and charging station efficiency is assumed to be 0.9. The values are retrieved from various literatures (Hota et al., 2014). Moreover, SoC is in the range of 0.2–0.8 (Chang, 2013).

3.3 METAHEURISTIC FRAMEWORK

3.3.1 Particle Swarm Optimization (PSO)

PSO is an evolutionary computation technique which is proposed by Eberhart and Yuhui (2001). The PSO is inspired from social behavior of birds flocking. It uses a number of particles (candidate solutions) which fly around in the search space to find the best solution. Meanwhile, they all look at the best particle (best solution) in their

TABLE 3.1
Parameter Settings of the Objective Function

Parameters	Values
Stable parameters	Maximum power, $P_{i,\max} = 6.7\,\text{kWh}$
	Charging station efficiency, $\eta = 0.9$
	Total charging time, $\Delta t = 20$ min
	Power allocation to each PHEV: 30W
Variables	$0.2 \le$ state-of-charge (SoC) ≤ 0.8
	Waiting time ≤ 30 min (1800 s)
	16 kWh $\le$ battery capacity (C_i) $\le$ 40 kWh
Constraints	$\sum_i P_i(k) \le P_{\text{utility}}(k) \times \eta$
	$0 \le P_i(k) \le P_{i,\max}(k)$
	$0 \le \text{SoC}_i(k) \le \text{SoC}_{i,\max}$
	$0 \le \text{SoC}_i(k+1) - \text{SoC}_i(k) \le \Delta\text{SoC}_{\max}$

paths. In other words, particles consider their own best solutions as well as the best solution that has been found so far.

A PSO system begins with a primary initial population of random individuals and signifies solutions of problems, to which are allocated random velocities. Each particle in a PSO should consider the current position, the current velocity, the distance to pbest, and the distance to gbest to modify its position. PSO was mathematically modeled as follows:

$$V_i^{t+1} = \omega v_i^t + c_1 \cdot \text{rand} \cdot \left(\text{pbest} - x_i^t\right) + c_2 \cdot \text{rand} \cdot \left(\text{gbest} - x_i^t\right) \tag{3.11}$$

$$x_i^{t+1} = x_i^t + V_i^{t+1} \tag{3.12}$$

Generally speaking, the principal steps in PSO can be summarized as follows:

1. Generate a group of random solutions (particles) in the feasible region. Since we normally have very little information about the global optima, these particles are scattered over the search space as uniformly as possible.
2. Evaluate the distance between the new solution and the desired solution based on a fitness function.
3. Compare the fitness value at the current iteration with previous best, and update the individual best (pbest) and group best (gbest).
4. Update the velocities of the particles.
5. Update the positions of the particles.
6. Repeat steps 2–5 until the maximum number of iterations or the minimum error criteria are met.

Here v_i^t is the velocity of particle i at iteration t and w is a weighting function usually used as follows:

$$\omega - \omega_{max} - \frac{w_{max} - \omega_{min}}{Itre_{max}} Itre \quad (3.13)$$

3.3.1.1 Selecting PSO Parameters

Appropriate values for ω_{min} and w_{max} are 0.4 and 0.9. Appropriate value range for c_1 and c_2 is 1 to 2, but 2 is most appropriate in many cases. rand is a random number between 0 and 1, x_i^t is the current position of particle i at iteration t, pbest$_i$ is the pbest of agent i at iteration t, and gbest is the best solution so far. The parameter settings for PSO are demonstrated in Table 3.2. Total size of the swarm is 100 and PSO inertia is taken as 0.9. PSO is also fairly immune to the size and nonlinear nature of the objective function being considered. The algorithm does not converge with less iterations, while more iterations increase computation complexity; so the maximum iterations are 100.

The main advantage of PSO is its simplicity, while being capable of delivering accurate results consistently. It is fast and also very flexible, being applicable to a wide range of problems, with limited computational requirements (Eberhart & Yuhui, 2001). For these reasons, the present work focuses on metaheuristics optimization approaches, namely PSO applied in order to optimize the SoC for charging PHEVs.

The system initially has a population of random selective solutions. Each potential solution is called a particle. Each particle is given a random velocity and is flown through the problem space. The particles have memory and each particle keeps track of its previous best position (called the pbest) and its corresponding fitness. There exist a number of pbest for the respective particles in the swarm and the particle with greatest fitness is called the global best (gbest) of the swarm. The basic concept of the PSO technique lies in accelerating each particle toward its pbest and gbest locations, with a random weighted acceleration at each time step (Ganesan, Vasant, & Elamvazuthy, 2012).

TABLE 3.2
PSO Parameter Settings

Parameters	Values
Size of the swarm	100
Maximum number of steps	100
PSO parameter, c_1	1.4
PSO parameter, c_2	1.4
PSO inertia (w)	0.9
Maximum iterations	100
Number of runs	50

Algorithm 1: Particle Swarm Optimization

Step 1: Particle's position and velocity initialization.
Step 2: Particle's velocity update.
Step 3: Particle's position update in two stages.
Step 4: Particle's best-known position "pbest" update.
Step 5: Swarm's best-known position "gbest" update.
Step 6: Set new population = current population.
Step 7: If the termination conditions are satisfied halt and print solutions, else go to step 2.

3.3.2 Accelerated Particle Swarm Optimization (APSO)

In APSO, each member of the population is called a particle and the population is called a swarm. Starting with a randomly initialized population and moving in randomly chosen directions, each particle moves through the searching space and remembers the best earlier positions, velocity, and accelerations of itself and its neighbors. Particles of a swarm communicate good position, velocity, and acceleration to each other as well as dynamically adjust their own position, velocity, and acceleration derived from the best position of all particles. The next step starts when all particles have been shifted. Finally, all particles are inclined to fly toward better positions over the searching process until the swarm moves close to an optimum of the fitness function.

The standard PSO uses both the current global best and the pbest. The reason of using the pbest is mainly to increase the diversity in the quality solutions; however, this diversity can be simulated using some randomness. Subsequently, there is no convincing reason for using the pbest, unless the optimization problem of interest is multimodal and highly nonlinear.

It is worth pointing out that there is no need to deal with initialization of velocity vectors. Therefore, the APSO is much simpler. Compared with many PSO variants, the APSO uses only two parameters, and the mechanism is simple to understand. In APSO, for the optimization we have considered three parameters—position, velocity, and acceleration—for each swarm particle, whereas in PSO only two parameters—position and velocity—are considered for each particle. Here in this algorithm the swarms are in the random sequence and rand positions are generated. From these positions, the velocity and acceleration are generated.

The outline of APSO is given below:

Algorithm 2: Accelerated Particle Swarm Optimization

Step 1: Generation of primary population.
Step 2: Random initialization of initial position and velocity.
Step 3: Evaluate particle velocity.
Step 4: Evaluate particle position in single step.

Step 5: Actual position of each particle.
Step 6: Set new population = current population.
Step 7: If the termination conditions are satisfied halt and print solutions, else go to step 2.

A simplified version which could accelerate the convergence of the algorithm is to use the global best only. Thus, in the APSO, the velocity vector is generated by a simpler formula where randn is drawn from (0,1) to replace the second term. The update of the position is simply like Equation 3.15.

$$V_i^{t+1} = V_i^t + \alpha \cdot \text{randn}(t) + \beta \cdot \left(g^* - x_i^t\right) \tag{3.14}$$

where randn is drawn from N (0,1) and the update of the position is like the standard PSO method. In order to increase the convergence even further, the update of the position can be written in a single step, as

$$x_i^{t+1} = (1-\beta)x_i^t + \beta \cdot g^* + \alpha r \tag{3.15}$$

3.3.2.1 Selecting APSO Parameters

In our simulation we use

$$\alpha = 0.7^t \tag{3.16}$$

The typical values for this accelerated PSO are $\alpha \approx 0.1$–0.4 and $\beta \approx 0.1$–0.7; however, $\alpha \approx 0.2$ and $\beta \approx 0.5$ are recommended (Mohamed et al., 2012). In general, any evolutionary search algorithm shows improved performance with a relatively larger population. However, a very large population will cost more in terms of fitness function evaluations without producing significant improvements. In this simulation, the population size is set to 100. Total 30 independent runs were performed with 100 iterations each time. The parameter settings for APSO are demonstrated in Table 3.3.

TABLE 3.3
APSO Parameter Settings

Parameters	Values
Size of the swarm	100
Maximum number of steps	100
Alpha, α	0.2
Beta, β	0.5
Maximum iterations	100
Number of runs	30

TABLE 3.4
Average Best Fitness for APSO and PSO

Computational Time (s)	PSO	APSO
50 PHEVs	1.650	1.685
100 PHEVs	1.686	1.690
500 PHEVs	1.990	1.856
1000 PHEVs	2.398	2.141

TABLE 3.5
Computational Time for APSO and PSO

Average Best Fitness for	PSO	APSO
50 PHEVs	142.839	165.96509
100 PHEVs	171.102	182.93134
500 PHEVs	150.869	197.59083
1000 PHEVs	156.802	172.45284

3.4 SIMULATION OUTCOMES AND ANALYSIS

The PSO and APSO algorithms were applied to find out global best fitness of the objective function. All the simulations were run on a Core™ i5-3470M CPU@ 3.20 GHz processor, 4.00 GB RAM and MATLAB R2013a.

Table 3.4 summarizes the simulation results for 50, 100, 500, and 1000 PHEVs, respectively, for finding the maximum fitness value of objective function $J(k)$. In order to evaluate the performance and show the efficiency and superiority of the proposed algorithm, we ran each scenario total 30 times. So it can be concluded that APSO outperformed PSO in terms of average best fitness. Starting from 50 numbers of PHEVs up to 1000 PHEVs, APSO shows better fitness value than PSO.

Table 3.5 shows the computational time requirement for APSO and PSO methods. As the number of PHEVs increased from 100 to 500 and 1000, APSO technique showed better result than standard PSO method in terms of computational time.

It can be apparently seen that although the algorithm has been set to run for maximum 100 iterations, the fitness value converges after 10 iterations and becomes stable. So, there is an early convergence which may cause the fitness function to trap into local minima. This can be avoided by increasing the size of swarm; hence the computational time will also be increased. As a result, a trade-off should be taken into consideration between the proper convergence and computational time (Rahman, Vasant, Singh, & Abdullah-Al-Wadud, 2014c).

3.5 COMPARISON BETWEEN PSO AND APSO

For solving this particular optimization problem, we faced some issues which will be discussed in this section.

Although APSO needs more parameter tuning compared to standard PSO method, when the number of PHEVs increases, APSO takes less time than PSO. This characteristic makes APSO very efficient to solve this particular optimization problem. In APSO, the velocity vector ensures local exploitation capability. Moreover, the disadvantage of APSO is that it suffers early convergences in primary stages.

The advantages of PSO algorithm:

1. The algorithm is simple, with less adjustable parameters and being easy to implement.
2. The random initialization of population has strong global search ability, which is similar to genetic algorithm (GA).
3. It uses the evaluation function to measure the individual searching speed.
4. It has strong scalability.

The disadvantages of PSO algorithm:

1. The algorithm cannot make full use of the feedback information in the system.
2. Its ability to solve combinatorial optimization problems is not strong.
3. It is easy for this algorithm to obtain local optimal solution.

The advantages of APSO algorithm:

1. Very efficient in terms of solving the particular fitness function.
2. The velocity vector is generated by a simpler formula.
3. Update of the position can be written in a single step.
4. Local exploitation ability to find good solutions.

The disadvantages of APSO algorithm:

1. Suffers from early convergence in the primary stages.
2. α and β parameters have effects on the algorithm performance and no fixed value can ensure higher fitness values.

Finally, from Figure 3.1, we come to a conclusion that APSO performs better than PSO in terms of average best fitness for up to 1000 PHEVs.

3.6 DISCUSSION

3.6.1 Computational Cost

For real-life glitches, the computational cost of a full evaluation of the fitness function can easily become the dominant computational cost. This computational cost can have the effect of making the time for the swarm to converge sluggishly. In this APSO method, the computational cost is moderate as compared to standard PSO method because of using acceleration factors, α and β.

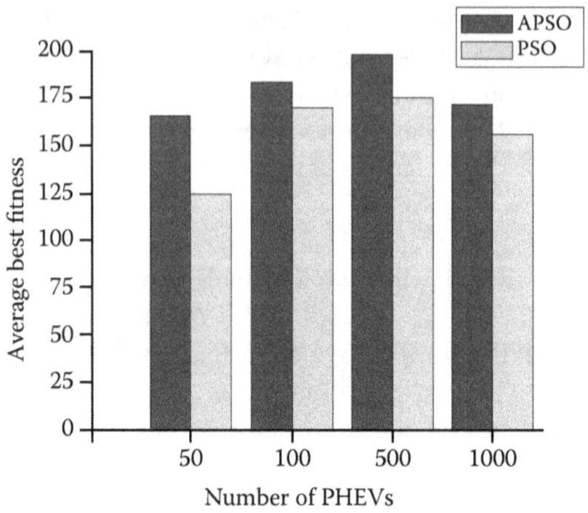

FIGURE 3.1 Average best fitness versus number of PHEVs (APSO and PSO).

3.6.2 Stopping Criteria

Since an iterative method computes successive approximations to the solution of a system, stopping criteria are needed to determine when to stop the iteration. The maximum number of iterations was set to 100 for this optimization.

3.6.3 Robustness

The robustness of the algorithm is tested in terms of the variability of the final solutions from each set of experiments. APSO algorithm is not too robust as the maximum, average, and minimum fitness values show different values for different number of PHEVs. Tuning the parameters such as α and β will improve the robustness of this optimization which is beyond the scope of this research.

3.6.4 Computational Complexity

Computational complexity refers to the different problems encountered in solving an optimization algorithm, such as premature convergence, high computational time, trapping in local optima, incapability of reaching global optima/minima, and so on. In our optimization problem, we encounter premature convergences. Moreover, if the size of swarm is very small, then the algorithm traps in local minima. In order to avoid this, we started our simulation using standard swarm size which is 100. In future, more swarm sizes will be considered in order to find a global solution.

3.7 FUTURE RESEARCH DIRECTIONS

This paragraph encapsulates the review results and suggests future directions of optimization techniques and procedures. The specific research field is relatively new and possible future perspectives have to be emphasized, so that new techniques can be realized.

3.7.1 Optimization Techniques

Optimization techniques such as evolutionary algorithms, direct search methods, and other heuristic methods should be introduced in order to avoid the calculation of function derivatives. Multiobjective capability should also be provided for multi-criteria optimization problems. The future optimization tools should be capable of performing parallel processing evaluations on the same computer by using modern multicore processor technology or to distribute the calculations to a cluster of computers. Such ability will substantially improve the simulation runtime. Advanced controlling mechanisms are necessary for allocating sufficient energy to a particular charging station in order to facilitate large-scale PHEV penetration in upcoming years. The future optimization tools should have the capability of stable convergence and thus provide a good solution to the desired fitness functions. Exploration and exploitation of the search space are essential in order to get desired solution within acceptable computation time. Finally, optimization of charging station needs proper assortment of available resources as well as efficient available technique implementation. Possible characteristics of the future optimization tools are given below.

3.7.1.1 Ant Colony Optimization (ACO)

Ant colony optimization (ACO), introduced by Dorigo in his doctoral dissertation, is a class of optimization algorithms modeled on the actions of an ant colony. ACO is a probabilistic technique useful in problems that deal with finding better paths through graphs. Artificial "ants" simulation agents locate optimal solutions by moving through a parameter space representing all possible solutions (Dorigo, 2006). Natural ants lay down pheromones directing each other to resources while exploring their environment. The simulated "ants" similarly record their positions and the quality of their solutions, so that in later simulation iterations more ants locate better solutions. For solving this highly nonlinear fitness function, ACO method can be applied in order to achieve high fitness value and also less computation time. This method is a member of swarm intelligence (SI) group. So, similar results are expected for PSO and APSO. Researchers should apply this method for solving charging problem of PHEVs and find the outcomes with extensive comparison with other swarm intelligence-based methods.

3.7.1.2 ABC Optimization

ABC algorithm is one of the most recently introduced swarm-based algorithms. ABC simulates the intelligent foraging behavior of a honeybee swarm (Karaboga & Basturk, 2007). In the ABC model, the colony consists of three groups of bees: employed bees, onlookers, and scouts. It is assumed that there is only one artificial employed bee for each food source. In other words, the number of employed bees in the colony is equal to the number of food sources around the hive. Employed bees go to their food source and come back to the hive and dance on this area. The employed bees whose food source has been abandoned become scouts and start searching for finding a new food source. Onlookers watch the dances of employed bees and choose food sources depending on dances. In ABC, a population-based algorithm, the position of a food source represents a possible solution to the optimization problem and the nectar amount of a food source corresponds to the quality (fitness) of the associated solution. The number of the employed bees is equal to the number of solutions in the population. Future studies for solving smart charging problem of PHEVs should involve the ABC optimization technique.

3.7.2 Demand Side Management

Demand side management (DSM) is defined by the Department of Energy (DOE) as, "changes in electric usage by end-use customers from their normal consumption patterns in response to changes in the price of electricity over time, or incentive payments designed to induce lower electricity use at times of high wholesale market prices or when system reliability is jeopardized." Therefore, the DSM programs should be incorporated into the existing Intelligent Energy Management System (iEMS) model in order to avoid voltage sag and blackout and to maximize the commercial benefits. In addition, this underutilized capacity could effectively power a national fleet of PHEVs with little need to increase the energy delivery capacity of the existing grid infrastructure (Gerkensmeyer, Kintner-Meyer, & DeSteese, 2010).

3.7.3 Trade-Off between Cost and Performance

Considering the effects of cost and performance on the marketability of PHEVs, the fitness function is defined to minimize drivetrain cost and driving performance requirements are selected as constraints to ensure that the vehicle performance is not sacrificed during the optimization. The battery is the key component within an EV which determines its overall capital cost and performance. Therefore, the task of determining the cost-effectiveness of EVs is predominantly one of identifying the future trajectory of battery cost and performance. To meet power requirements, batteries have lower discharge power at low SoC and lower charge power at high SoC. To reduce safety risks, limiting the maximum SoC avoids overcharge situations.

3.8 CONCLUSION AND RECOMMENDATIONS

In this chapter, APSO-based optimization was performed in order to allocate power optimally to each of the PHEVs entering into the charging station. A sophisticated controller will need to be designed in order to allocate power to PHEVs appropriately. For this wake, the applied algorithm in this chapter is a step toward real-life implementation of such a controller for PHEV charging infrastructures. Here, four different numbers of PHEVs were considered for MATLAB simulation and then obtained results were compared with PSO in terms of average best fitness and computational time. The success of the electrification of transportation sector solely depends on charging infrastructure. Only proper charging control and infrastructure management can assure the larger penetration of PHEVs. The researchers should try to develop efficient control mechanisms for charging infrastructure in order to facilitate upcoming PHEVs penetration in highways. In future, more vehicles should be considered for intelligent power allocation strategy as well as hybrid versions of PSO should be applied to ensure higher fitness value and low computational time.

REFERENCES

Boyle, G. (Ed.). (2007). *Renewable electricity and the grid: The challenge of variability.* London: Earthscan.

Caramanis, M., & Foster, J. M. (2009). Management of electric vehicle charging to mitigate renewable generation intermittency and distribution network congestion. In *Proceedings of the 48th IEEE Conference on Decision and Control,* 2009 December. Shanghai: IEEE.

Chang, W. Y. (2013). The state of charge estimating methods for battery: A review. *ISRN Applied Mathematics, 2013,* Article ID 953792, 7 pages. http://dx.doi.org/10.1155/2013/953792.

Chiasson, J., & Vairamohan, B. (2005). Estimating the state of charge of a battery. *IEEE Transactions on Control Systems Technology, 13*(3), 465–470.

del Valle, Y., Venayagamoorthy, G. K., Mohagheghi, S., Hernandez, J.-C., & Harley, R. G. (2008). Particle swarm optimization: Basic concepts, variants and applications in power systems. *Evolutionary Computation, IEEE Transactions, 12*(2), 171–195.

Dorigo, M. (Ed.). (2006). *Ant colony optimization and swarm intelligence: Proceedings of the 5th International Workshop, ANTS 2006, Brussels, Belgium,* September 4–7, 2006 (Vol. 4150). Berlin, Heidelberg: Springer-Verlag.

Eberhart, R. C., & Yuhui, S. (2001). Particle swarm optimization: Developments, applications and resources. In *Proceedings of the 2001 Congress on Evolutionary Computation.* Seoul, South Korea: IEEE.

Ganesan, T., Vasant, P., & Elamvazuthy, I. (2012). A hybrid PSO approach for solving nonconvex optimization problems. *Archives of Control Sciences, 22*(1), 87–105.

Gerkensmeyer, C., Kintner-Meyer, M. C., & DeSteese, J. G. (2010). *Technical challenges of plug-in hybrid electric vehicles and impacts to the US power system: Distribution system analysis.* Washington, DC: United States Department of Energy.

Hess, A., Malandrino, F., Reinhardt, M. B., Casetti, C., Hummel, K. A., & Barceló-Ordinas, J. M. (2012). Optimal deployment of charging stations for electric vehicular networks. In *Proceedings of the First Workshop on Urban Networking.* Nice, France: ACM.

Hota, A. R., Juvvanapudi, M., & Bajpai, P. (2014). Issues and solution approaches in PHEV integration to the smart grid. *Renewable and Sustainable Energy Reviews, 30*(0), 217–229.

Jia-Zhao, C., Yu-Xiang, Z., & Yin-Sheng, L. (2012). A unified frame of swarm intelligence optimization algorithm. *Knowledge Discovery and Data Mining, 135,* 745–751.

Karaboga, D., & Basturk, B. (2007). A powerful and efficient algorithm for numerical function optimization: Artificial bee colony (ABC) algorithm. *Journal of Global Optimization, 39*(3), 459–471.

Li, Z., Sahinoglu, Z., Tao, Z., & Teo, K. H. (2010). Electric vehicles network with nomadic portable charging stations. In *Vehicular Technology Conference* Fall, 2010, September. Ottawa, Canada: IEEE.

Lund, H., & Kempton, W. (2008). Integration of renewable energy into the transport and electricity sectors through V2G. *Energy Policy, 36*(9), 3578–3587.

Mayfield, D. (2012). *Site design for electric vehicle charging stations.* Washington, DC: US Department of Energy.

Mohamed, A. Z., Lee, S. H., Hsu, H. Y., & Nath, N. (2012). A faster path planner using accelerated particle swarm optimization. *Artificial Life and Robotics, 17*(2), 233–240.

Morrow, K., Karner, D., & Francfort, J. (2008). *Plug-in hybrid electric vehicle charging infrastructure review.* US Department of Energy-Vehicle Technologies Program.

Piller, S., Perrin, M., & Jossen, A. (2001). Methods for state-of-charge determination and their applications. *Journal of Power Sources, 96*(1), 113–120.

Prajna, K., Rao, G. S. B., Reddy, K., & Maheswari, R. U. (2014). A new dual channel speech enhancement approach based on accelerated particle swarm optimization (APSO). *International Journal of Intelligent Systems and Applications (IJISA), 6*(4), 1.

Rahman, I., Vasant, P. M., Singh, B. S. M., & Abdullah-Al-Wadud, M. (2014a). Intelligent energy allocation strategy for PHEV charging station using gravitational search algorithm. *AIP Conference Proceedings, 1621*, 52–59.

Rahman, I., Vasant, P. M., Singh, B. S. M., & Abdullah-Al-Wadud, M. (2014b). Optimisation of PHEV/EV charging infrastructures: A review. *International Journal of Energy Technology and Policy, 10*(3/4), 280–296.

Rahman, I., Vasant, P. M., Singh, B. S. M., & Abdullah-Al-Wadud, M. (2014c). Swarm intelligence-based smart energy allocation strategy for charging stations of plug-in hybrid electric vehicles. *Mathematical Problems in Engineering, 2014*, Article ID 620425, 10 pages, 2014. doi:10.1155/2014/620425.

Soares, J., Sousa, T., Morais, H., Vale, Z., Canizes, B., & Silva, A. (2013). Application specific modified particle swarm optimization for energy resource scheduling considering vehicle-to-grid. *Applied Soft Computing, 13*(11), 4264–4280.

Su, W., & Chow, M. Y. (2011). Performance evaluation of a PHEV parking station using particle swarm optimization. In *Power and Energy Society General Meeting,* 2011 IEEE (pp. 1–6), July. IEEE.

Su, W., & Chow, M.-Y. (2012a). Performance evaluation of an EDA-based large-scale plug-in hybrid electric vehicle charging algorithm. *Smart Grid, IEEE Transactions on, 3*(1), 308–315.

Su, W., & Chow, M.-Y. (2012b). Computational intelligence-based energy management for a large-scale PHEV/PEV enabled municipal parking deck. *Applied Energy, 96*, 171–182.

Tie, S. F., & Tan, C. W. (2013). A review of energy sources and energy management system in electric vehicles. *Renewable and Sustainable Energy Reviews, 20*, 82–102.

Part II

Process Optimization

4 Three-Phase Separator Optimization Using Bacteria Foraging

In the oil and gas industry, one of the most important stages in processing petroleum is separation. It can be classified by operating configurations such as vertical, horizontal, and spherical or by its function, which is two phase or three phase. In this chapter, the vertical three-phase separator was chosen and researched. A three-phase separator is used to separate water, oil, and gas. Gas forms at the top, oil forms at the middle layer, and water forms at the bottom due to gravitational force and the density of the substance. The objective was to tune the proportional–integral–derivative (PID) controller controlling the level of the water in the separator. Outflow rate of the water from the bottom of the separator will be used to control the water level. Currently, there are controlling methods, namely Proportional Integral (PI) control using trial-and-error method, PI control using Butterworth filter design method, and internal model control (IMC) method. These methods were having quite high percent overshoot and long settling time. Therefore, this chapter discusses the use of bacterial foraging optimization algorithm (BFOA) in optimizing the parameters for PI control of a three-phase separator.

4.1 INTRODUCTION

A three-phase separator is used to separate water, oil, and gas where the gas forms at the top, oil forms at the middle layer, and water forms at the bottom due to gravitational force and the density of the substance.

PID control is the most commonly used control algorithm in a three-phase separator. There are a few methods in selecting the proper parameters which are manual method, Ziegler–Nichols method, and Cohen–Coon tuning method. In the manual method, experienced personnel have to select the optimum parameters by trial and error as well as by observing the output response of the system. Ziegler–Nichols tuning method requires trial and error to obtain the ultimate gain and period. Then, some minor calculation will be done in selecting the parameters. Cohen–Coon tuning method only applies to first-order processes with dead time and a series of calculations need to be done in selecting the parameters. In terms of three-phase separator, the conventional methods used are IMC method, PI controller using Butterworth filter design method, and PI controller using trial-and-error method. All these methods require time as well as cost and the performance of the PID controller might not be satisfactory.

Currently, there are intelligent algorithms that can select and optimize the parameters of the plant, for example: genetic algorithm (GA), particle swarm optimization (PSO), ant colony optimization (ACO), and BFOA. These techniques find and optimize the parameters faster than the conventional tuning methods and are very cost-effective.

Over the past years, BFOA method has been applied in many industrial fields. In this context, PID controller design for a power plant (Bensenouci, 2011), a new power system stabilizer design (Daryabeigi, Moazzami, Khodabakhshian, & Mazidi, 2011), automated tuning of PID controllers of unmanned aerial vehicles (UAVs) (Oyekan & Hu, 2010), bacterial foraging oriented by PSO strategy for PID tuning (Korani, Dorrah, & Emara, 2009), high-dimensional function optimization (Chu, Mi, Liao, Ji, & Wu, 2008), optimization design of PID controller parameters based on improved *Escherichia coli* foraging optimization algorithm (Yijian & Yanjun, 2008), improved bacterial foraging strategy for controller optimization applied to robotic manipulator system (Coelho & Silveira, 2006), and optimum design of PID controllers using only a germ of intelligence (Niu, Zhu, He, & Zeng, 2006), are just a few of the applications.

In this chapter, the details of the mathematical model of three-phase separator, the working principle of BFOA, simulation of BFOA and three-phase separator, and performance comparison of BFOA with other existing techniques are presented. The scope of the study takes into consideration the percentage overshoot, settling time, and integral squared error (ISE) after implementing BFOA of the PID controller of different orders and applications.

4.2 THREE-PHASE SEPARATOR

4.2.1 Operation of Three-Phase Separator

A three-phase separator is created to separate and remove water from the mixture of water and crude oil. Figure 4.1 is a schematic diagram of a horizontal three-phase separator.

The fluid will hit the inlet diverter after entering the separator. There form an initial gross separation of the liquid and vapor due to the sudden change in momentum

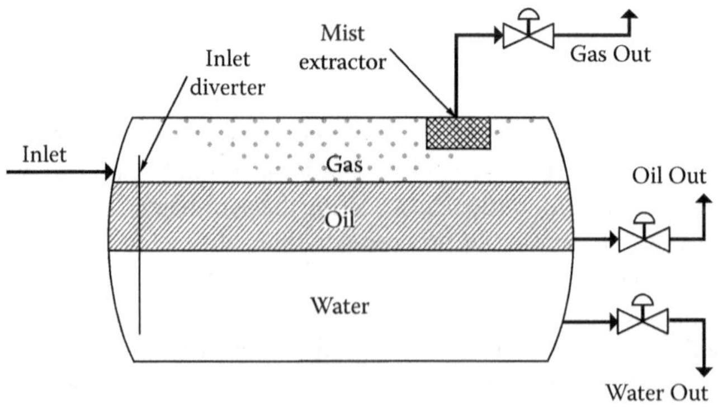

FIGURE 4.1 Schematic diagram of a horizontal three-phase separator.

of the fluid. This forces the inlet mixture of oil and water to mix with the water continuous phase in the bottom of the vessel and rise to the oil–water interface. This process is called "water-washing"; it promotes the coalescence of water droplets which are entrained in the oil continuous phase. Some of the gas flows over the inlet diverter and then horizontally through the gravity settling section above the liquid. As the gas flows through this section, small drops of liquid that were entrained in the gas and not separated by the inlet diverter are separated out by gravity and fall to the gas–liquid interface. Some of the drops are of such a small diameter that they are not easily separated in the gravity settling section. Before the gas leaves the vessel it passes through a coalescing section or mist extractor to coalesce and remove them before the gas leaves the vessel (Sayda & Taylor, 2007).

When the hydrocarbon fluid stream enters a three-phase separator, two distinctive phenomena take place. The first phenomenon is fluid dynamic, which is characterized by the gravity separation of oil and water droplets entrained in the aqueous and the oil phases respectively, the gravity separation of gas bubbles entrained in the stream, and the gravity separation of liquid droplets which are dispersed in the gas phase. The second phenomenon is thermodynamic, in the sense that some light hydrocarbons and gas solution flash out the oil phase and reach a state of equilibrium due to the pressure drop in the separator. Furthermore, preceding gravity separation processes minimize the amounts of gas entrained in the main stream. Figure 4.2 shows the simplified separation process. Oil-well fluid with molar flow F_{in} and gas, oil, and water molar fractions Z_g, Z_o, and Z_w, respectively, enters the separator.

The hydrocarbon component of the fluid separates into two parts; the first stream F_{h1} separates by gravity and enters the oil phase, and the second stream F_{h2} stays in the aqueous phase due to incomplete separation. The gas component in the separated hydrocarbon stream, which enters the oil phase, separates into two parts: the first gas stream F_{g1} flashes out of the oil phase due to the pressure drop in the separator, and the second gas stream F_{g2} stays dissolved in the oil phase. The oil discharge $F_{o_{out}}$ from the separator contains the oil component of the separated hydrocarbon F_o and the dissolved gas component F_{g2}. The flashed gas $F_{g_{out}}$ flows out of the separator for further processing.

The water and oil levels and the gas pressure inside the separator are controlled by a number of separate control systems. Figure 4.3 shows the control loop involved in a three-phase separator.

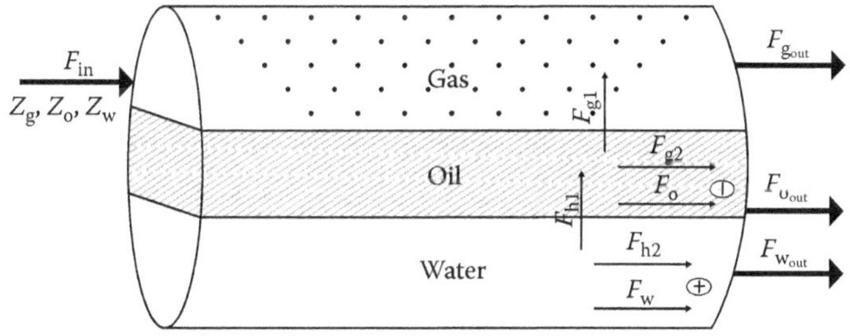

FIGURE 4.2 Simplified separation process. (Adapted from Sayda, A.F. and Taylor, J.H. Modeling and control of three-phase gravity separators in oil production facilities. *American Control Conference*, New York City, July 2007.)

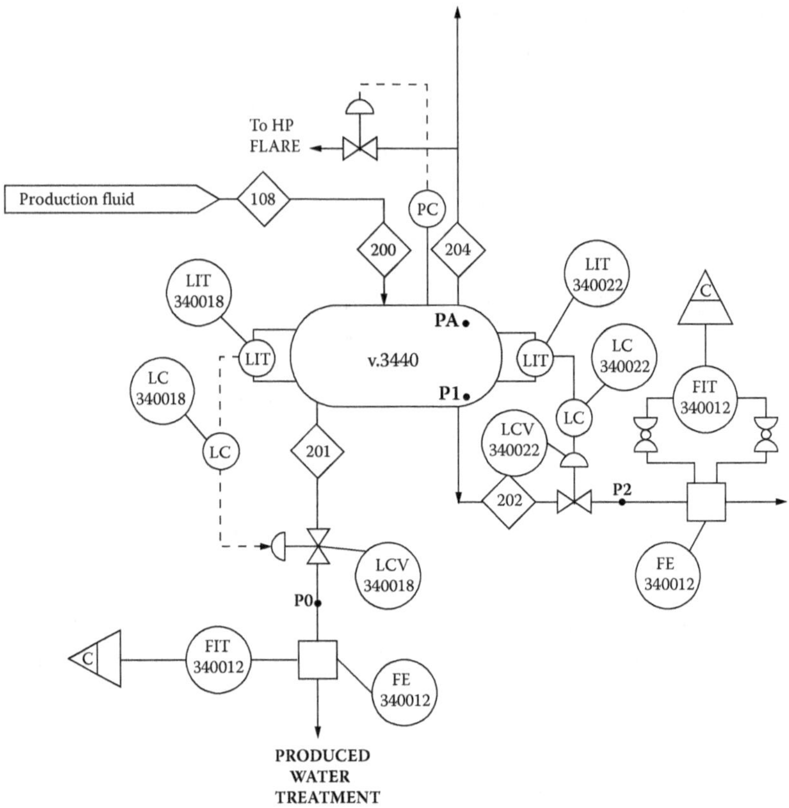

FIGURE 4.3 P&I diagram of relevant control loop of three-phase separator. (Adapted from Yang, Z., et al. On the innovation of level control of an offshore three-phase separator. Department of Electronic Systems, Aalborg University, Ramboll Oil and Gas A/S, Denmark, 2010.)

It can be observed that a level indicator transmitter (tagged LIT-340018) is employed to measure the water level inside the separator if we focus on the water-level control loop. The measured level signal is sent to a level controller, tagged LC-340018. The level controller sends the control signal to a level control valve, tagged LCV-340018. In order to control the water level inside the separator, the LCV-340018 regulates the water outflow. It can be noticed that a flow indicator transmitter, named FIT-340012, is used to measure the water outflow rate for some other purpose. This measurement is not used by the current level controller (Yang et al., 2010).

4.3 BACTERIA FORAGING OPTIMIZATION ALGORITHM (BFOA)

4.3.1 Behavior of Bacteria Foraging

Considering the behavior of the bacteria in searching for nutrients, BFOA mimicked it by focusing on the four main sections which are chemotaxis, swarming, reproduction as well as elimination, and dispersal.

4.3.1.1 Chemotaxis

Chemotaxis is the movement of the bacteria in search of food or nutrients by taking small steps. The bacteria tumble around randomly until they find higher concentration or gradient of nutrient. Then, the bacteria will run toward that direction until they reach the highest food concentration.

This process simulates the movement of an *E. coli* cell through swimming and tumbling via flagella. Biologically, an *E. coli* bacterium can move in two different ways. It can swim for a period in the same direction or it may tumble, and alternate between these two modes of operation for the entire lifetime.

Suppose $\theta^i(j+1, k, l)$ represents i-th bacterium at j-th chemotactic, k-th reproductive, and l-th elimination–dispersal step. $C(i)$ is the size of the step taken in the random direction specified by the tumble (run length unit). Then in computational chemotaxis the movement of the bacterium may be represented by

$$\theta^i(j+1,k,l)=\theta^i(j,k,l)+C(i)\frac{\Delta(i)}{\sqrt{\Delta^T(i)\Delta(i)}} \tag{4.1}$$

where Δ indicates a vector in the random direction whose elements lie in [–1, 1] (Das, Biswas, Dasgupta, & Abraham, 2009).

4.3.1.2 Swarming

Swarming is the behavior or the tendency of the bacteria to group together. Each bacterium will release signals to other bacteria to attract or to repel them.

An interesting group of behavior has been observed for several motile species of bacteria including *E. coli* and *Salmonella typhimurium*, where intricate and stable spatiotemporal patterns (swarms) are formed in semisolid nutrient medium. A group of *E. coli* cells arrange themselves in a traveling ring by moving up the nutrient gradient when placed amid a semisolid matrix with a single nutrient chemo-effecter. The cells when stimulated by a high level of *succinate* release an attractant *aspartate*, which helps them to aggregate into groups and thus move as concentric patterns of swarms with high bacterial density. The cell-to-cell signaling in *E. coli* swarm may be represented by the following function.

$$\begin{aligned} J_{cc}(\theta,P(j,k,l)) &= \sum_{i=1}^{S} J_{cc}(\theta,\theta^i(j,k,l)) \\ &= \sum_{i=1}^{S}\left[-d_{\text{attractant}}\exp\left(-w_{\text{attractant}}\sum_{m=1}^{p}(\theta_m-\theta_m^i)^2\right)\right] \\ &\quad+\sum_{i=1}^{S}\left[-h_{\text{repellant}}\exp\left(-w_{\text{repellant}}\sum_{m=1}^{p}(\theta_m-\theta_m^i)^2\right)\right] \end{aligned} \tag{4.2}$$

where $J_{cc}(\theta, P(j, k, l))$ is the objective function value to be added to the actual objective function (to be minimized) to present a time-varying objective function, S is the total number of bacteria, p is the number of variables to be optimized, which are present in each bacterium, and $\theta = \left[\theta_1, \theta_2, \ldots, \theta_p\right]^{\mathrm{T}}$ is a point in the p-dimensional search domain. $d_{\text{attractant}}$, $w_{\text{attractant}}$, $h_{\text{repellant}}$, and $w_{\text{repellant}}$ are different coefficients that should be chosen properly (Das et al., 2009; Korani et al., 2009; Bensenouci, 2011).

4.3.1.3 Reproduction

Reproduction is the behavior of bacteria where the weaker half of the group of bacteria will die and the other stronger half will split and reproduce asexually. The number of bacteria of the group will remain constant.

For the given k and l, and for each $i = 1, 2, \ldots, S$, let

$$J_{\text{health}}^{i} = \sum_{j=1}^{N_c+1} J(i, j, k, l) \tag{4.3}$$

be the health of the bacterium i (a measure of how many nutrients it got over its lifetime and how successful it was at avoiding noxious substances). Sort bacteria and chemotactic parameters $C(i)$ in order of ascending cost J_{health} (higher cost means lower health).

4.3.1.4 Elimination and Dispersal

Elimination and dispersal are the probability of a bacterium being eliminated in a group and if elimination has happened, another bacterium will be dispersed to a random location in the optimization domain.

4.3.1.5 Combination of All Four Parts

For each bacterium, the cost of its movements involved the swarming effect from the other bacteria and the chemotaxis effect. The equations had accounted for a whole set of chemotactic steps for each swarming effect done. The equation of the swarming effect is represented by

$$J(i, j+1, k, l) = J(i, j, k, l) + J_{\text{cc}}\left(\theta^i(j, k, l), P(j, k, l)\right) \tag{4.4}$$

Then, for each swarming effect, a set of chemotactic steps will be calculated. A chemotactic step is represented by

$$\theta^i(j+1, k, l) = \theta^i(j, k, l) + C(i)\frac{\Delta(i)}{\sqrt{\Delta^{\mathrm{T}}(i)\Delta(i)}} \tag{4.5}$$

and the cost or $J(i, j, k, l)$ can be calculated from that equation. A series of $J(i, j, k, l)$ will be calculated and at the same time the new value will be compared with the previous one. The lower value or cost will be the better one. Then, another bacterium will be selected, the swarming effect as well as the chemotactic steps will be recalculated, and the process will be repeated for every other bacteria. Besides, this bacterial movement will be repeated for N_c times in random direction to obtain the best results. After N_c times of looping, the $J(i, j, k, l)$ for each bacterium will be summed up to obtain J_{health}. J_{health} will be sorted, the weaker half will die, the other half will reproduce asexually, and the whole process will start over again. Elimination and dispersal process will eliminate a bacterium and disperse another one to its domain with a probability of P_{ed} after N_{re} times of reproduction.

4.3.2 Relation with PID Controller

The bacteria undergo chemotaxis, swarming, reproduction as well as elimination, and dispersal in order to survive and maintain the population by considering the optimum concentration of food, tendency of swarming, and the health of the bacteria. Hence, the cost finding the food and maintaining in a group has to be low. In other words, bacteria use the smallest amount of effort to achieve best results. The formulas or methods mentioned previously are combined and show the behavior of the bacteria.

By implementing this survival technique in searching for the best parameters, ISE which is represented by J_{health} (higher cost means lower health) will be chosen. BFOA will implement convergence on the parameter values to the optimum level and produce the best result in a short time. So, the optimum parameter of K_p, K_i, and K_d will be chosen.

4.4 IMPLEMENTATION

4.4.1 Plant with PID Controller

Figure 4.4 is the block diagram of a plant where the algorithm will take the input and find the optimum parameters which are K_p, K_i, and K_d. Then, PID controller will take the parameters to control the system.

4.4.2 Flow of BFOA

Figure 4.5 shows the flowchart of the BFOA.

It can be seen from Figure 4.5 that it represents all the four parts, that is, chemotaxis, swarming, reproduction as well as elimination, and dispersal. J_{last} with the smallest value will be produced and chosen since it represents the ISE and is also known as cost.

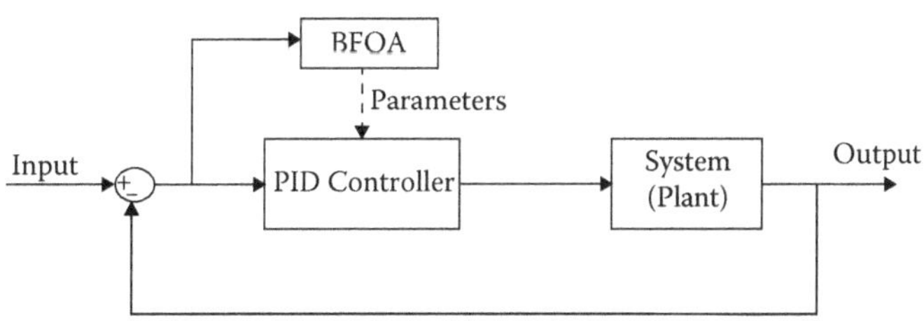

FIGURE 4.4 Block diagram of the system with PID controller and BFOA.

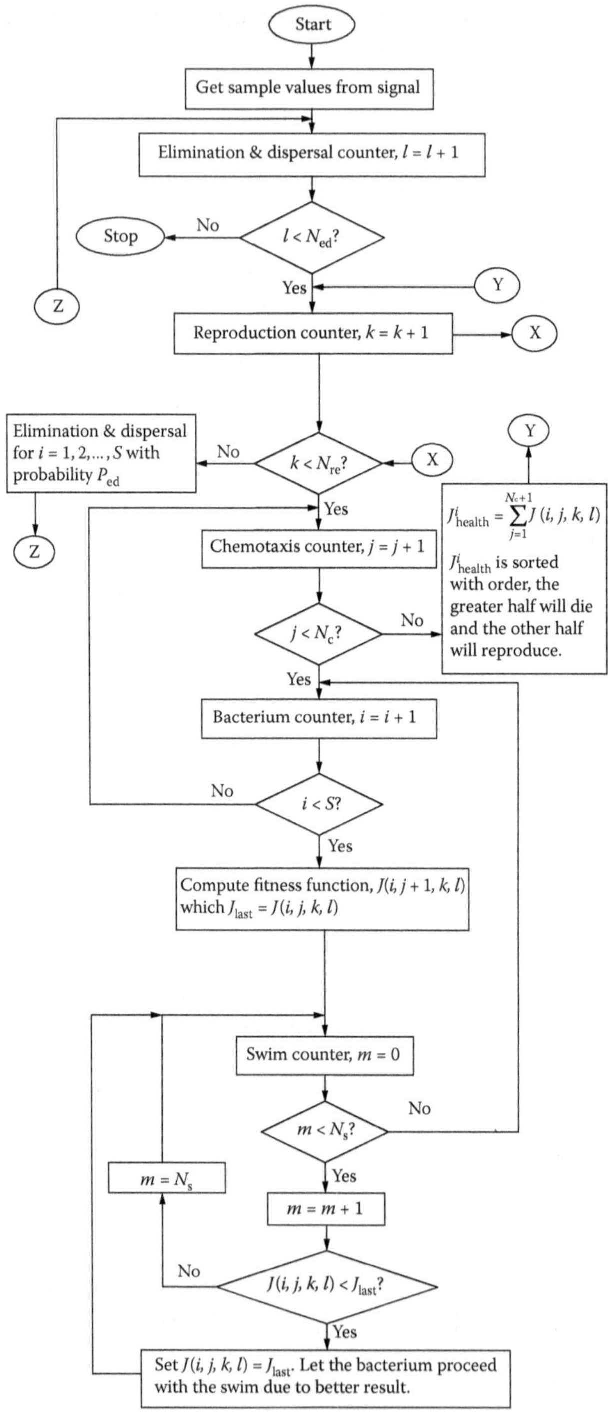

FIGURE 4.5 Flowchart of the bacterial foraging optimization algorithm (BFOA).

4.4.3 Mathematical Model of Three-Phase Separator

A three-phase separator can be modeled and represented by using a transfer function to show its characteristics. Figure 4.6 shows the block diagram of three-phase separator with disturbance, with Q_{in} included.

BFOA is used and implemented as shown in Figure 4.6 on top of *C*(*s*). The input, *R*(*s*), and output, *H*(*s*), are the desired level and actual level of the water in the separator.

$$G_1(s)=\frac{H(s)}{Q_{in}(s)}=\frac{1}{47.55s+1.81} \tag{4.6a}$$

$$G_2(s)=\frac{H(s)}{U(s)}=-\frac{10.82}{47.55s+1.81} \tag{4.6b}$$

Equations 4.6a and b show the transfer function of the plant and the disturbance (Yang et al., 2010).

4.4.4 Simulation

For each of the methods, the parameters that were used are shown in Table 4.1.

And for IMC method (Yang et al., 2010),

$$\text{Internal model control}=-\frac{47.55s+1.81}{45.958s} \tag{4.7}$$

By using MATLAB® Simulink®, the following system was constructed as shown in Figure 4.7.

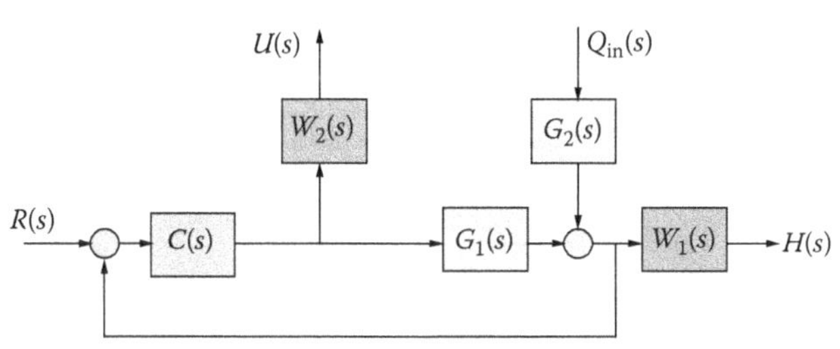

FIGURE 4.6 Block diagram of three-phase separator with disturbance.

TABLE 4.1
Values of Parameters for Each Method

Method	K_p	K_i
BFOA	−33.898	−0.1065
Trial and error[a]	−0.7391	−1.582
Butterworth[a]	−1.05	−1.76

[a] Yang et al. (2010).

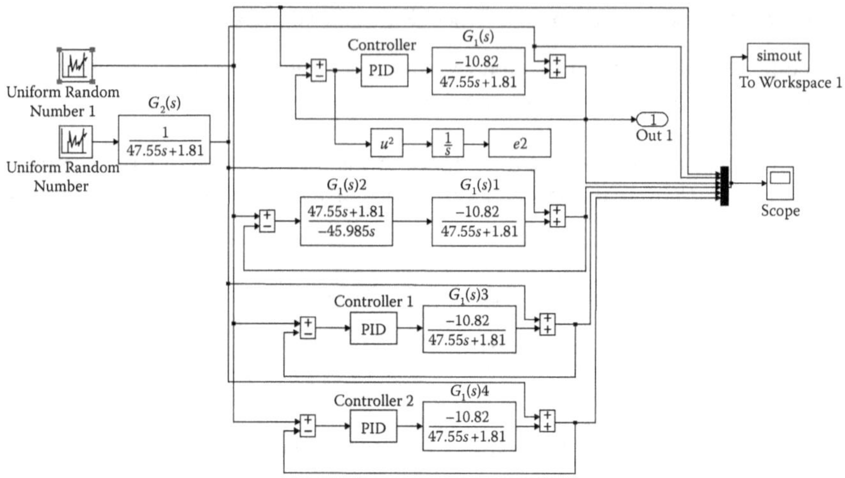

FIGURE 4.7 MATLAB Simulink of block diagram comparing all methods.

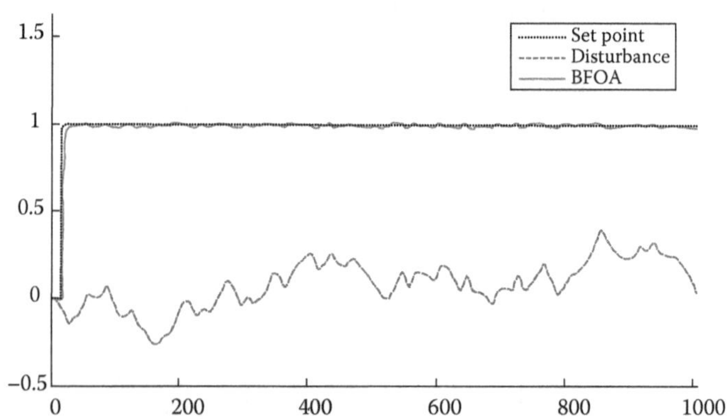

FIGURE 4.8 Response of plant using BFOA with disturbance.

4.5 RESULTS AND ANALYSIS

The results obtained for each of the methods through the simulation are illustrated in Figures 4.8 through 4.15. Figures 4.8 through 4.11 show the initial set-point change followed by the disturbance due to the input well fluids of the three-phase separator.

Figures 4.12 through 4.15 show the responses which include the input disturbance and the set point changes as well.

Figures 4.8 through 4.11 show that the response of using BFOA tuning method produces the best results which are almost the same as the desired results. Trial-and-error method and Butterworth filter design method produce results that have high overshoots and undershoots. IMC method produces slow responses which have long rise time. IMC method, trial-and-error method, and Butterworth filter design method have large settling time or very slow responses compared to BFOA tuning method.

Figures 4.16 through 4.18 show the ISE, percentage overshoot, and settling time respectively for all methods.

The lower the values, the better the response is. For these three figures, BFOA shows the lowest values. This has proved that BFOA optimizes better parameters compared with these conventional methods. From all of the four methods used, BFOA shows the most significant improvements by having the least or almost 0% overshoot and the shortest settling time. This had shown that BFOA performs the best of all the methods tested.

Besides comparing the results of BFOA and the conventional methods, a Graphical User Interface (GUI) was developed to ease the user in optimizing the parameters using BFOA. The software developed can optimize parameters for PID or PI controllers with up to third-order system. Figure 4.19 is a screen shot of the GUI of the BFOA based software:

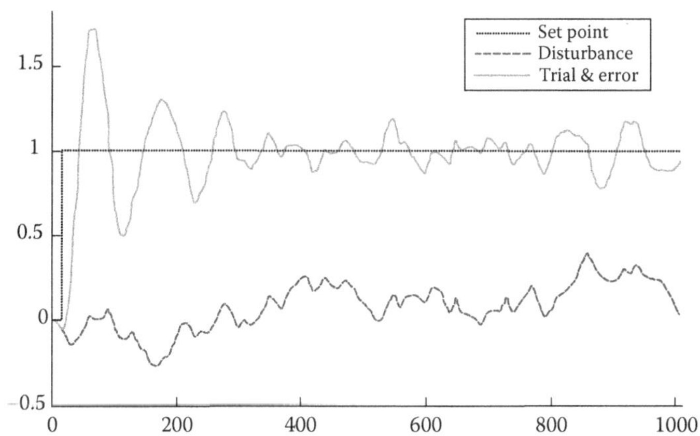

FIGURE 4.9 Response of plant using trial and error with disturbance.

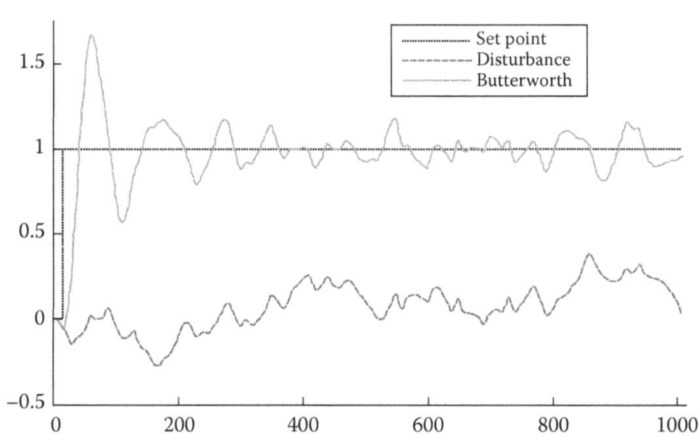

FIGURE 4.10 Response of plant using Butterworth filter design method with disturbance.

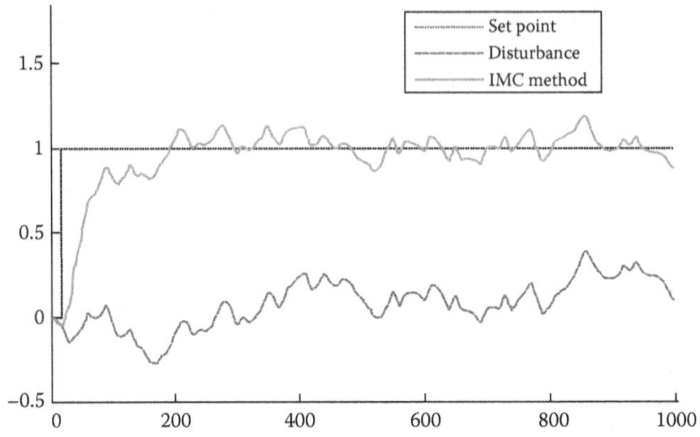

FIGURE 4.11 Response of plant using IMC method with disturbance.

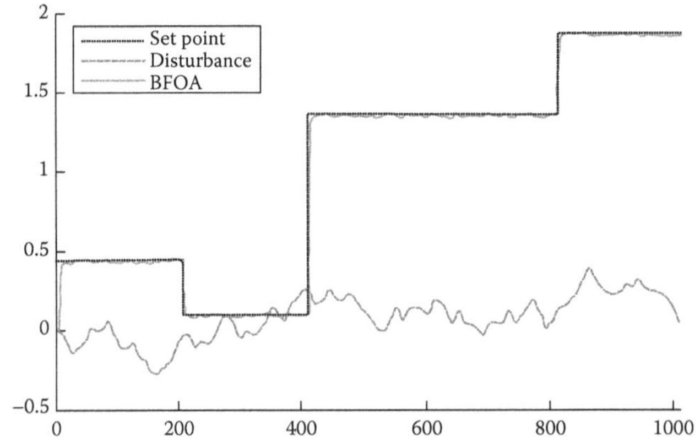

FIGURE 4.12 Response of plant using BFOA with disturbance and set-point changes.

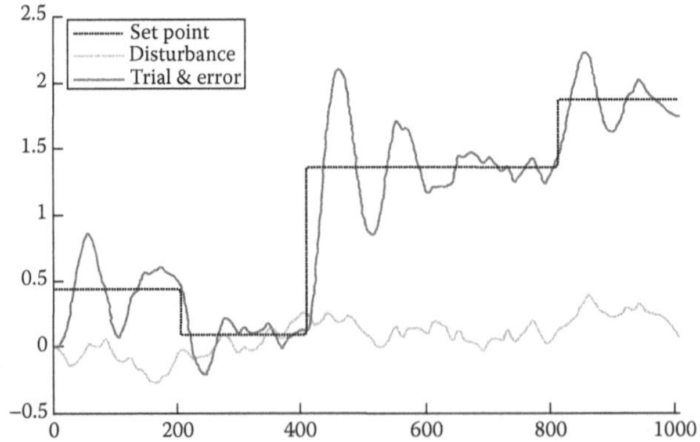

FIGURE 4.13 Response of plant using trial and error with disturbance and set-point changes.

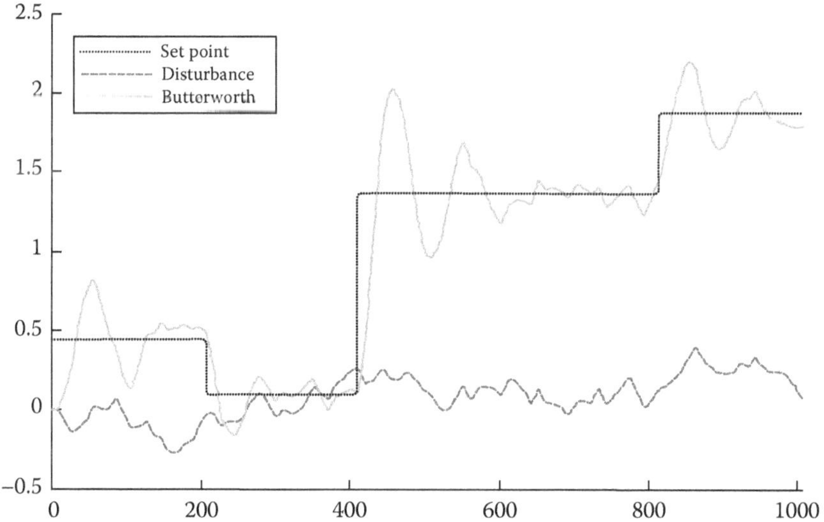

FIGURE 4.14 Response of plant using Butterworth filter design with disturbance and set-point changes.

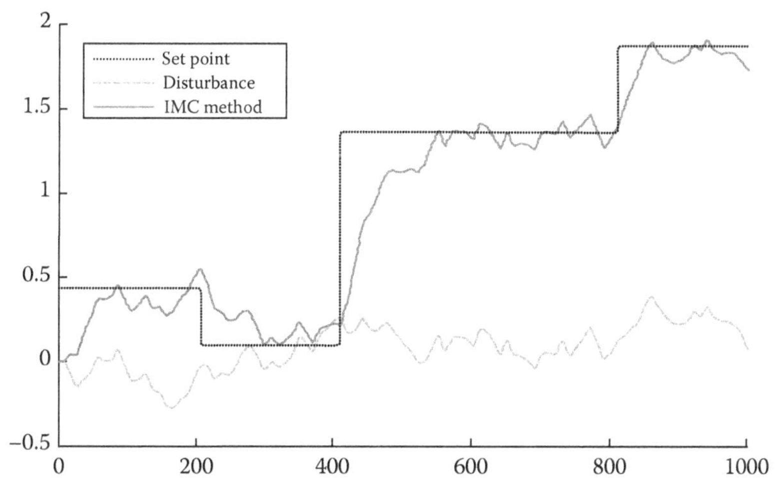

FIGURE 4.15 Response of plant using IMC method with disturbance and set point changes.

Transfer function can be input into the GUI in the first box and then generate the equation to initialize the required parameters. Then, optimization can be done. After that, values of K_p, K_i, and K_d will be displayed and a step response of the system can be plotted.

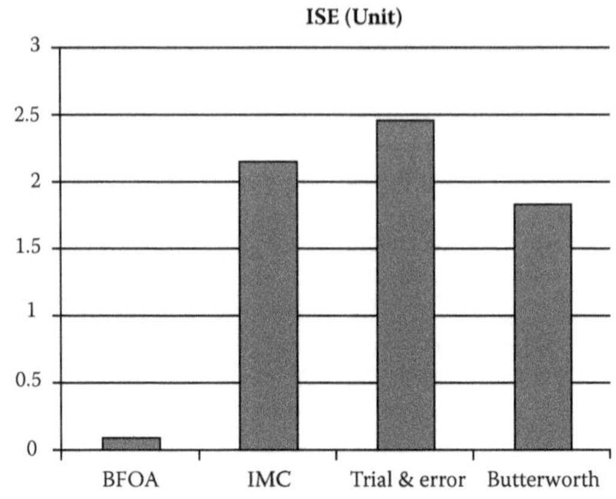

FIGURE 4.16 Integral squared error of different methods.

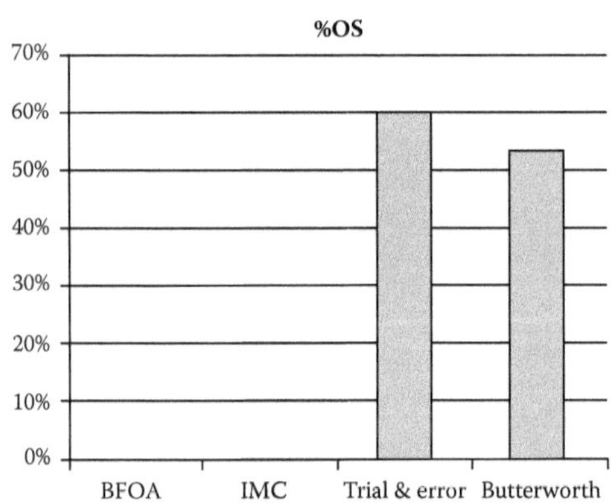

FIGURE 4.17 Percentage overshoot of different methods.

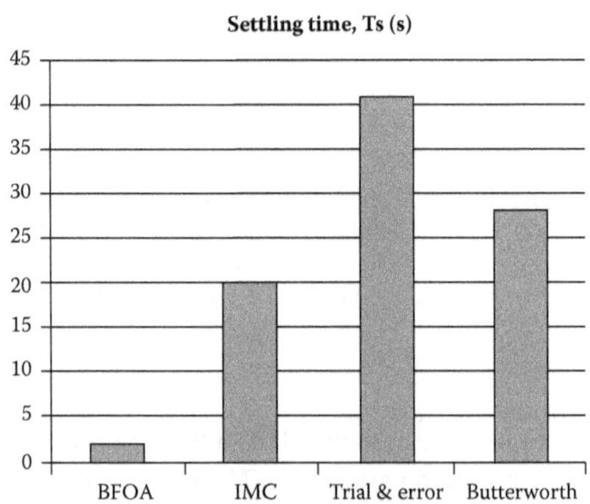

FIGURE 4.18 Percentage overshoot of different methods.

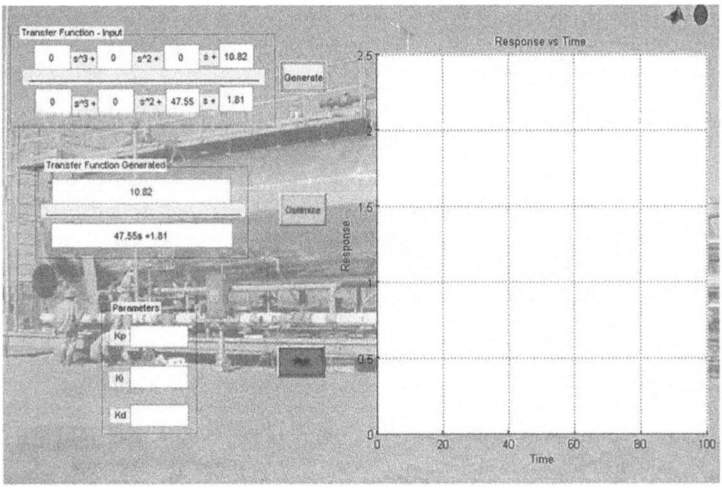

FIGURE 4.19 Graphical user interface (GUI) for BFOA based software.

4.6 SUMMARY

A mathematical model of the three-phase separator was derived and based on the model, trial-and-error method, Butterworth filter design method, IMC method, and BFOA tuning method were tested using MATLAB Simulink simulation. A set of parameters was used and the results were obtained. The results have shown that the BFOA method produces the best results and far better that the other three conventional methods. BFOA has more than 90% improvements compared to the conventional methods. BFOA can improve the performance of the system and thus increase productivity, decrease cost, and save time. A GUI is developed in order to ease the optimization process. This is a promising technique to be used in oil and gas industry other than three-phase separator. Further studies can be carried out on an actual plant and plants with higher order. Besides, BFOA can be used to compare with other intelligent techniques.

REFERENCES

Bensenouci, A. (2011). PID controllers design for a power plant using bacteria foraging algorithm, Electrical and Computer Engineering Department, King Abdulaziz University, Jeddah, Kingdom of Saudi Arabia, 2011.

Chu, Y., Mi, H., Liao, H., Ji, Z., & Wu, Q. H. (2008). A fast bacterial swarming algorithm for high-dimensional function optimization, IEEE Congress on Evolutionary Computation (CEC 2008), Hong Kong, China, 2008.

Coelho, L. S., & Silveira, C. C. (2006). Improved bacterial foraging strategy for controller optimization applied to robotic manipulator system, Pontifical Catholic University of Parana, Brazil, October 4–6, 2006.

Daryabeigi, E., Moazzami, M., Khodabakhshian, A., & Mazidi, M. H. (2011). A new power system stabilizer design by using smart bacteria foraging algorithm, Department of Electrical Engineering, Islamic Azad University, Iran, 2011.

Das, S., Biswas, A., Dasgupta, S., & Abraham, A. (2009). Bacterial foraging optimization algorithm: Theoretical foundations, analysis, and applications, *Foundations of Computational Intelligence Volume 3*, Volume 203 of the series Studies in Computational Intelligence (pp. 23–55). Springer, http://www.springer.com/gp/book/9783642010842?wt_mc=ThirdParty.SpringerLink.3.EPR653.About_eBook.

Korani, W. M., Dorrah, H. T., & Emara, H. M., (2009). Bacterial foraging oriented by particle swarm optimization strategy for PID tuning. *Proceedings of the IEEE International Symposium on Computational Intelligence in Robotics and Automation*, December 15–18, 2009. Daejeon, South Korea (pp. 445–450).

Niu, B., Zhu, Y., He, X., & Zeng, X. (2006). Optimum design of PID controllers using only a germ of intelligence, Shenyang Institute of Automation, Chinese Academy of Sciences, Liaoning, China, June 21–23, 2006.

Oyekan, J., & Hu, H. (2010). A novel bacterial foraging algorithm for automated tuning of PID controllers of UAVs, School of Computer Science and Electronic Engineering, University of Essex, Wivenhoe Park, Colchester, UK June 10–23, 2010.

Sayda, A. F., & Taylor, J. H. (2007). Modeling and control of three-phase gravity separators in oil production facilities, *American Control Conference*, New York City, July 2007.

Yang, Z., Juhl, M., & Lohndorf, B. (2010). On the innovation of level control of an offshore three-phase separator. Department of Electronic Systems, Aalborg University, Ramboll Oil and Gas A/S, Denmark.

Yijian, L., & Yanjun, F. (2008). Optimization design of PID controller parameters based on improved *E. coli* foraging optimization algorithm, School of Electrical & Automation Engineering, Nanjing Normal University, Department of Automation, University of Wuhan, September 2008.

5 Plant Optimization Using Artificial Fish Swarm Algorithm

Optimization has always been the goal in engineering applications but the traditional optimization techniques are incapable of solving many of the problems because they often depend on the quality of the objective functions and they need a vast amount of computation. In this chapter, the artificial fish swarm algorithm (AFSA) is proposed to solve the PID controller tuning problem for several process plants. Although the Ziegler–Nichols (ZN) tuning method and other related methods are the workhorse of the industrial applications, this chapter shows that the AFSA is able to produce better performance compared to ZN method and other swarm intelligence techniques such as particle swarm optimization (PSO) method.

5.1 INTRODUCTION

The three-term proportional–integral–derivative (PID) controller is playing a vital role in the field of automation and control, and has been looked upon as a reliable component of industry because of its simplicity and satisfactory performance with a vast range of processes. Because of its modest structure, it has the ability to be easily understood and implemented in real-world environment. Thus, its presence is highly appreciated in practical applications. Conventionally, the PID parameters are tuned by the most commonly used ZN method. But these conventional tuning methods have several limitations (Astrom & Hang, 1991; Visioli, 2006).

Therefore, recently, many intelligent approaches are used. Of these, swarm approaches are gaining popularity. In engineering studies, the terminology of "swarm" is defined as a group of agents having independent individual dynamics but exhibiting closely coupled behaviors and performing any specific task as a group (Gazi & Passino, 2011). Another terminology that relates to such a system is termed as "multi-agent dynamics system" (Luo, Zhang, & Li, 2007). From a biological perspective, the term swarm relates to some species when they are in certain behavioral modes. Similar to swarm there are also some biological synonyms such as colonies of ants, flocks of birds, schools of fish, and so on. For many years, scientists and researchers in biology and engineering have been attracted to these coordinated behaviors of such clusters of species. A type of artificial intelligence (AI), swarm intelligence is the study of collective behavior of a reorganized system, usually a set of species that communicate locally among themselves as well as with the environment. There are many algorithms that use swarm techniques such as AFSA, ant colony optimization (ACO), PSO, honeybees optimization (HBO), and others (Li, Shao, & Qian, 2002; Luo, Wei, & Wang, 2010).

The AFSA is a recent and easy-to-implement artificial life-computing algorithm that simulates fish swarm behaviors and has been successfully used in some engineering applications. A variety of techniques have been proposed to handle the equality and inequality constraints of the problem. In this context, parameter estimation in engineering systems (Li, Xue, Lu, & Tian, 2004), determination of cluster number using average information entropy and density function based on fuzzy C-means cluster analysis and FSA (Zhu, Jiang, Song, & Bao, 2011), feed-forward neural networks (Wang, Zhou, & Ma, 2005), infinite impulse response (IIR) digital filter design (Yin & Tian, 2006), combinatorial optimization problem (Cai, 2010), forecasting stock indices using radial basis function neural networks (Shen, Guo, Wu, & Wu, 2011), Augmented Lagrangian fish swarm-based method for global optimization (Rocha, Martins, & Fernandes, 2011), hybridization of the FSA with the particle swarm algorithm to solve engineering systems (Tsai & Lin, 2011), combinatorial problems (Zheng & Lin, 2012), parameter estimation in control problems (Lobato, Souza, & Gedraite, 2012), and other applications (Wang, Gao, Cai, & Huang, 2006; Jiang, Wang, Pfletschinger, Lagunas, & Yuan, 2007).

In this chapter, the details on principles of AFSA and their implementation details for PID tuning are discussed. In addition, its application in four different process plants is explored. A comparative analysis of the performance of AFSA with ZN tuning and PSO methods was carried out.

5.2 ARTIFICIAL FISH SWARM ALGORITHM

5.2.1 Principles

In the underwater world, fish can find areas with more food based on their individual or swarm search. Inspired by this characteristic, the AFSA model is represented by searching, swarming, and following behaviors. The artificial fish (AF) model searches the problem space by those behaviors. The environment where the AF lives in is the problem space. The objective function of AFSA is to find maximum food density (Reza, 2014). Figure 5.1 shows that the AF model observes external concepts with its visual perception.

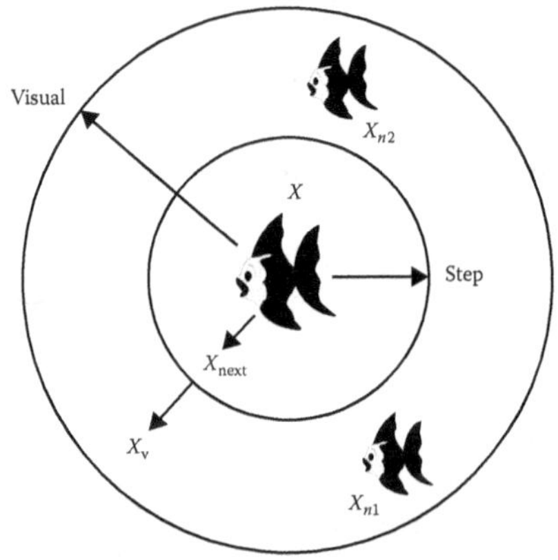

FIGURE 5.1 AFSA and its environment. (Adapted from Reza, A. *Int. J. Comput. Commun. Netw.*, 3, 1–7, 2014.)

5.2.1.1 Structure of AFSA

The current position of AF is shown by vector,

$$X = (x_1, x_2, \ldots, x_n) \tag{5.1}$$

where x_i $(i = 1, \ldots, n)$.

The Visual is equal to visibility domain of AF and X_v is an intended position in Visual where the AF selects to move toward. If X_v has better food density than current position, AF moves one step toward X_v, which results in displacement of the AF from X to X_{next}. Otherwise, if the current position of the AF is better than X_v, it selects another position in its Visual. Food density in position X is the fitness value of the position and is shown with Food Concentration (FC).

The state of AFSA individual can be represented as a vector; the variable X_i is for searching optimum value.

$FC = f(X)$, represents the present position of AF food consistence where the objective function is FC.

The total distance between two AF individuals is represented by:

$$d_{i,j} = \|X_i - X_j\| \tag{5.2}$$

where *Vision* is distance of visibility, *Step* is step length (maximum), and *Delta* (δ) is crowded factor.

AFSA model has got the most significance in the overall optimization process of PID parameters using AFSA. In the study the parameters that need to be optimized are K_p, K_i, and K_d. The AF vector can be expressed as

$$X = (K_p, T_i, T_d) \tag{5.3}$$

And the total distance between two AF individuals can be represented as

$$d_{i,j} = \sqrt{\left(K_p^i - K_p^j\right)^2 + \left(T_i^i - T_i^j\right)^2 + \left(T_d^i - T_d^j\right)^2} \tag{5.4}$$

At the initial position, the consistence of AF food can be represented by the steps outlined in following section.

5.2.1.2 Behavior Description

A. *Searching Action*

X_i = AF at initial position
X_j = AF in the visibility range

If $FC_i < FC_j$, proceed with a step within the same direction or else select X_j again.

$$x_{inextk} = x_{ik} + \text{Random(Step)}\frac{x_{jk} - x_{ik}}{\|X_j - X_i\|}, \ FC_j > FC_i \tag{5.5}$$

$$x_{inextk} = x_{ik} + \text{Random(Step)}, \ FC_j \le FC_i \tag{5.6}$$

where $K = 1, 2, 3, \ldots, n$; X_{jk}, X_{ik}, and X_{inextk} represent the k element of state vector X_j, X_i.

B. Swarming Action

X_c = AF state at center, FC_c = food availability at center.

n_f = number of fishes nearby ($d_{ij} \leq$ Visual), if $n_f \geq 1$, the center position of its fellow is discovered by X_j.

$$x_{ck} = \frac{\left(\sum_{j=1}^{n_f} x_{jk}\right)}{n_f} \tag{5.7}$$

$$x_{\text{inextk}} = x_{ik} + \text{Random(Step)}\frac{x_{ck} - x_{ik}}{\|X_c - X_i\|} \tag{5.8}$$

C. Following Action

X_i = AF at initial position

X_{max} = Max state

FC_{max} = largest amount of fellow fishes in nearby field ($d_{ij} \leq$ Visual)

If

$$\frac{FC_{max}}{\|n_f\|} > \delta FC_i \tag{5.9}$$

Fellow X_i has highest consistence of food and if the surrounding is not packed, then proceed with a step to Max.

$$x_{\text{inextk}} = x_{ik} + \text{Random(Step)}\frac{x_{\max k} - x_{ik}}{\|X_{max} - X_i\|} \tag{5.10}$$

If $n_f = 0$, AF individual performs the action of searching food.

The algorithm establishes a bulletin board and its purpose it to save the optimal position of AFSA and food consistence of the AFSA at corresponding position.

D. Effect of Parameters While Convergence

1. Visual and Step Searching and moving behavior is more prominent when the vision scope is smaller but in case of larger scope of vision the behavior of following is prominent. So for the better convergence larger scope of vision would be preferable, while the larger step is good for the convergence.

2. Crowd Factor (δ) For maximum desired value, the area of crowded factor is $0 < \delta < 1$. And, for minimal value, $\delta < 1$.

3. The Total of Fish Individual (N) If the number of fish is high the size of local extreme would be strong, leading to a faster convergence speed but the amount of iterations will be more.

5.2.2 PID Parameter Optimization Using AFSA

The essence of AFSA optimization searching is the estimation of fitness function. And its design is directly proportional to the performance of AFSA. This study is aimed at the parameters K_p, K_i, and K_d. To establish the objective function is a must to do for evaluation of performance index for searching and optimizing in a certain rule. The essence for designing PID controller is making minimum the system performance of index function J. In this study, the ITAE is used as performance index.

$$J(\text{ITAE}) = w\int_0^{\infty} t|e(t)|dt \tag{5.11}$$

The representation of objective function for this study is as follows:

$$J(\text{ITAE}) = \text{DT}^2 \sum_{k=1}^{\text{LP}} k|e(k)| \tag{5.12}$$

where DT is the calculation of step, and LP is the number of calculation.

In this study the searching is aimed to be maximum; so the fitness functions are fixed accordingly. Hence, the reciprocal of system performance index is used as fitness function.

$$\text{FC} = 1/J(\text{ITAE}) \tag{5.13}$$

AFSA creates a group of initialization of parameters as the fish colony to narrow down the area of PID parameters that are being optimized. These three parameters $\left(K_p^*, T_i^*, T_d^*\right)$ are first tuned by ZN approach and the center is expanded to both directions of polarities as follows:

$$\begin{aligned}(1-\lambda)K_p^* &\le K_p \le (1+\lambda)K_p^* \\ (1-\lambda)T_i^* &\le T_i \le (1+\lambda)T_i^* \\ (1-\lambda)T_d^* &\le T_d \le (1+\lambda)T_d^*\end{aligned} \tag{5.14}$$

where λ is the value in [1,0].

In AFSA the convergence is considered to terminate the algorithm when there is no higher value of fitness function after a number of iterations of searching.

5.2.3 Implementation Details

The plant parameters (K_p, K_i, and K_d) are tuned using AFSA as shown in Figures 5.2 and 5.3. The MATLAB Simulink Model was used to carry out the simulation.

Step 1: Randomly initialize the number of fish, the area of visibility, δ, and enter any random position of AF, $X = (x_1, x_2, \ldots, x_n)$.

Step 2: Calculate the food contamination of the defined position of AF. Obtain the concentration of food and updated position of X_i. The data must be saved in the bulletin box.

Step 3: For every AF, simulation reviews the results of swarming behavior and the following behavior. The behavior is chosen by each AF toward higher concentration of food. The searching behavior is by default behavior of the algorithm.

Step 4: Calculate the food contamination of all the AF at initial position as well as at each updated position. And then again the data must be saved in the bulletin box.

Step 5: If the iteration is equated to the limit, the optimal concentration of food and the corresponding position is the output.

else go to Step 3.

FIGURE 5.2 The AFSA scheme.

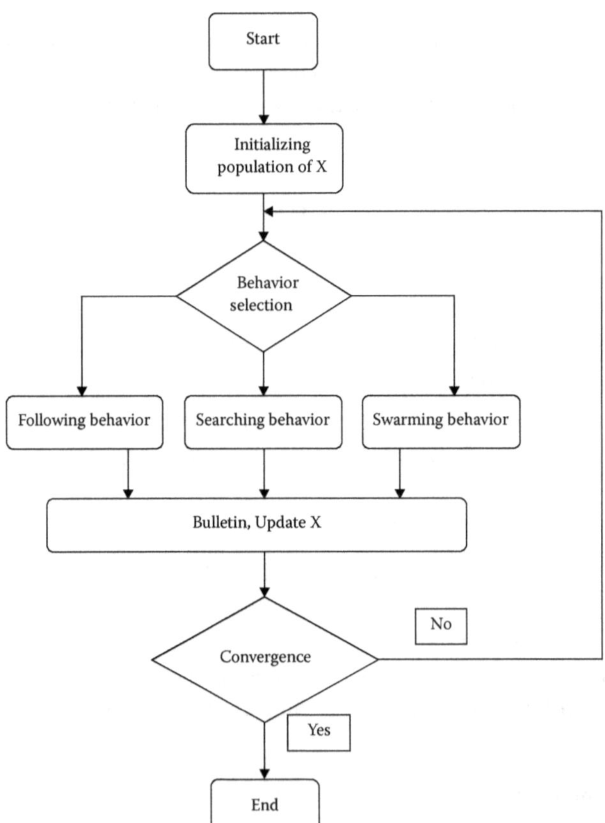

FIGURE 5.3 AFSA flowchart.

5.3 EXPERIMENTAL PLANTS

In industrial environments, typically, there are several plants that widely use PID tuning to optimize the output. In this study, five plants were studied. Pressure is one of the important conditions to ensure chemical reactions occur at a desired rate. It is also closely related to process temperature and both are closely monitored by plant operators to ensure safety in operation and maintaining product quality. As an example, the vapor-phase flow, pressure, and temperature plant's piping and instrumentation diagram (P&ID) diagram is shown in Figure 5.4, and is illustrated in Figure 5.5.

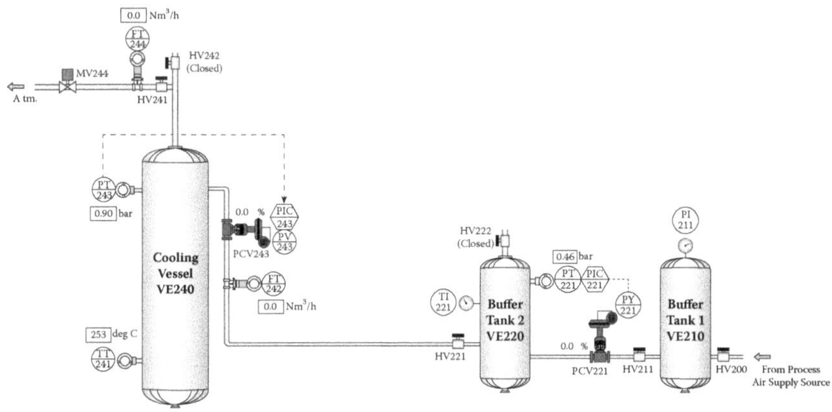

FIGURE 5.4 P&ID diagram of the scaled-down vapor-phase flow, pressure, and temperature process pilot plant.

FIGURE 5.5 Illustration of the scaled-down vapor-phase flow, pressure, and temperature process pilot plant.

The scaled-down vapor-phase flow, pressure, and temperature process pilot plant is a self-contained unit designed to simulate real flow, pressure, and temperature processes of a compressible fluid found in industrial plants Azhar (2010). Compressed air is used as the medium. This plant consists of two buffer tanks, *VE-210* and *VE-220*, which supply regulated, compressed air to the cooling vessel, *VE-240*. As more and more air enters *VE-240*, the inside pressure builds up. However, the compressed air is vented out at a certain rate through the solenoid valve *MV-244* to ensure a safe pressure limit inside *VE-240*.

5.4 FINDINGS, ANALYSIS, AND DISCUSSION

5.4.1 Modeling Based on System Identification

In order to get the transfer function of the plant, the process reaction curve is needed. The process reaction curve is shown in Figure 5.6 as indicated by the light gray line.

Table 5.1 shows the results obtained from process reaction curve.

The general transfer function of the plant is given as

$$G(s) = \frac{e^{-\theta s}}{\tau s + 1} \tag{5.15}$$

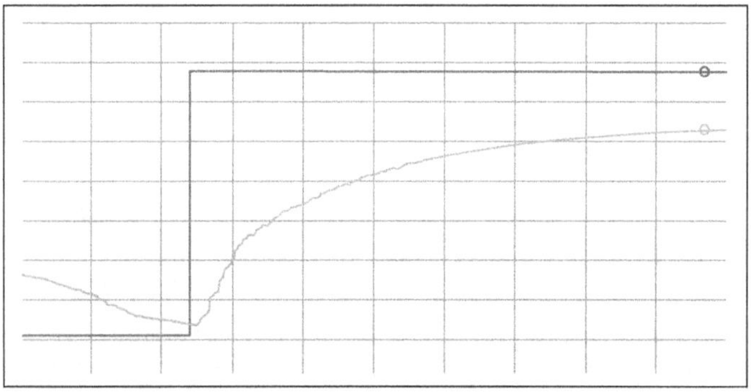

FIGURE 5.6 Process reaction curve.

TABLE 5.1
Readings from the Plant

Parameters	Values
Change in perturbation/MV, σ	0.1
Change in output/PV, Δ	0.21
Maximum slope, s	0.015
Apparent dead time, θ	11
Apparent time constant, $\tau = 1.5(0.63\Delta - 0.28\Delta)$	140

The derived transfer function of the plant based on the parameters of Table 5.1 is given as

$$G(s) = \frac{e^{-11s}}{140s+1} \tag{5.16}$$

5.4.2 ZN PID Tuning

The PID parameters that were obtained using the ZN method are shown in Table 5.2.

Initially, the PID controller is tuned by ZN method based on the parameter shown in Table 5.2. The simulated effect of the PID controller is shown in Figure 5.7.

5.4.3 Optimization of PID Parameters Using AFSA

The result of the AFSA simulation is shown in Figure 5.8.

Based on Figure 5.8, the new optimized parameters are shown in Table 5.3.

TABLE 5.2
PID Parameters

Controller	K_p	K_i	K_d
P	6.2	∞	0
PI	5.4	31.4	0
PID	8.19	26.208	3.943

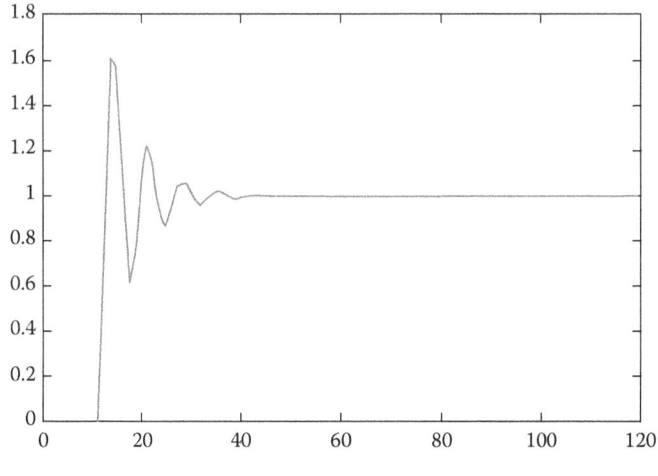

FIGURE 5.7 PID controller response tuned by Ziegler–Nichols method.

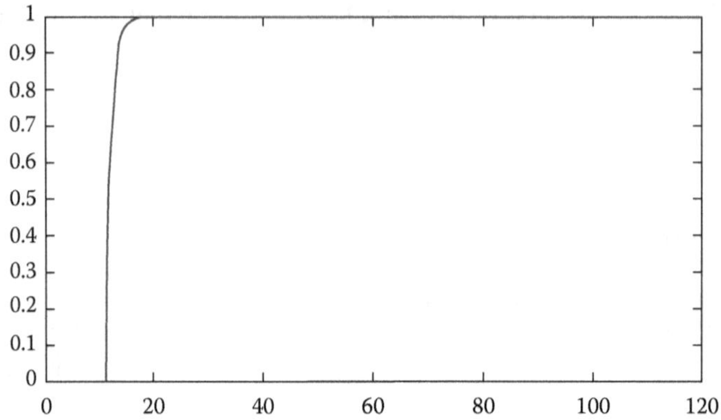

FIGURE 5.8 PID controller response tuned by AFSA method.

TABLE 5.3
AFSA Optimized Parameters

Controller	K_p	K_i	K_d	Percent Overshoot (%OS)	T_s
PID	36.57	0.0022	7.8492	0	20s

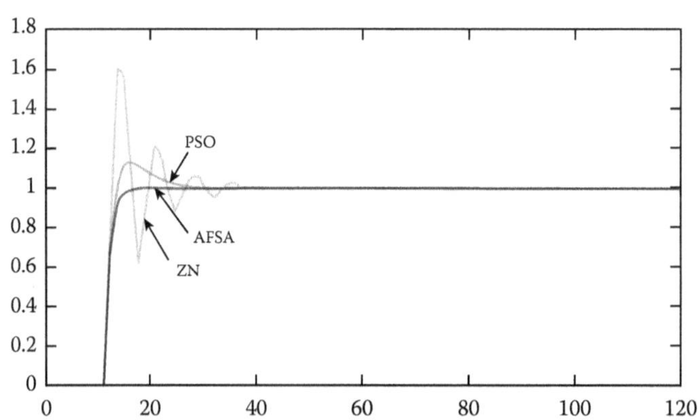

FIGURE 5.9 System responses of different optimization approaches.

5.4.4 Analysis and Discussion

It can be seen from Figures 5.7 and 5.8 that the system response using AFSA method was better than that using ZN method. The study was compared with other optimization methods such as PSO to validate and observe the mutual response of different optimization approaches. Figure 5.9 shows the responses of different optimization methods.

Based on Figure 5.9, the new optimized parameters are shown in Table 5.4.

TABLE 5.4
Performance Based on Different Optimization Approaches

Type	K_p	K_i	K_d	% OS	t_s
AFSA	36.57	0.0022	7.8492	0	20s
PSO	14.99	1772	13.872	15	30s
ZN-PID	8.1969	26.208	3.943	63	40s

Observing the optimization of PID parameters using different approaches in Table 5.4 we find that the intelligent methods present very acceptable results over the conventional methods of PID tuning. The traditional ZN method gives 63% of overshoot and requires some settling time, while PSO provides acceptably better results than ZN method with the overshoot of only 15% and less settling time. It can be seen that the PID parameter optimization using AFSA provides better results than both ZN and PSO methods where the overshoot is almost equal to 0% and the settling time is really less.

For the further validation of the study, an additional three plants were optimized using AFSA and compared with other optimization approaches as follows:

$$G(s) = \frac{0.5e^{-0.5s}}{1.24s^2 + 3.5s + 1} \tag{5.17}$$

$$G(s) = \frac{1}{s^2 + 0.1s + 1} \tag{5.18}$$

$$G(s) = \frac{e^{-0.5s}}{(s+1)^2} \tag{5.19}$$

The optimization results of the three plants are shown in Figures 5.10 through 5.12 and Tables 5.5 and 5.6.

The results from Figures 5.10 through 5.12 and Tables 5.5 through 5.7 show that for all the three plants, AFSA produces comparatively lower overshoots then other approaches, and the settling time is also reduced as compared to the other methods. Overall, the results show that the plant tuned with AFSA is superior in performance in terms of tuned parameters and percent overshoot.

5.5 SUMMARY

This chapter has shown that AFSA is an optimization technique that works effectively for the PID controller tuning problems. AFSA that was applied to four different plants in this study has shown superior performance compared to the traditional ZN and the PSO methods. It can be deduced that the AFSA has a major advantage, that is, it enables automatic tuning of PID parameters to produce system response of process plants. This would be a boon for engineers if it is implemented in the real-world environment.

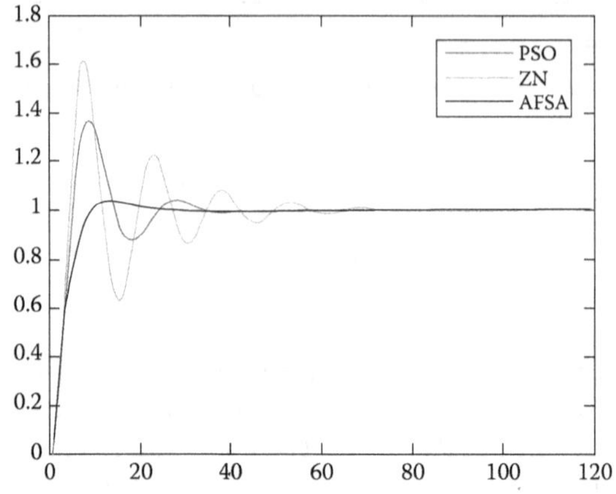

FIGURE 5.10 System response of plant I.

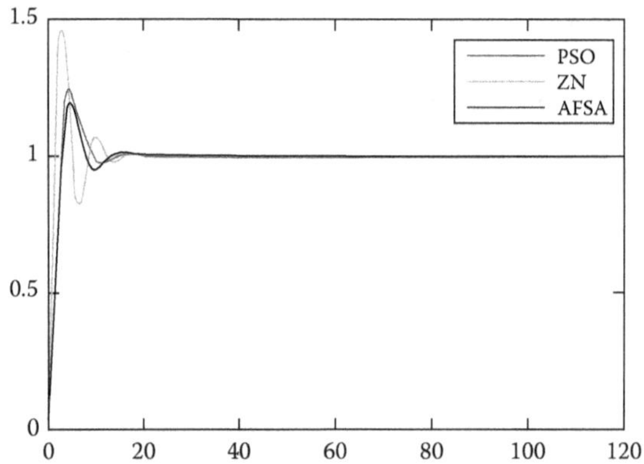

FIGURE 5.11 System response of plant II.

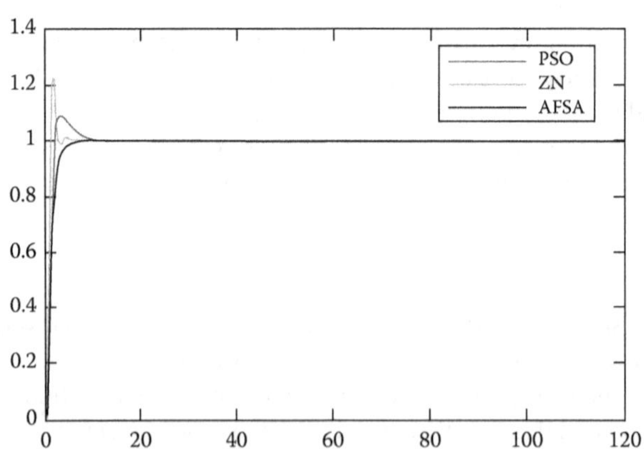

FIGURE 5.12 System response of plant III.

TABLE 5.5
Performance Based on Different Optimization Approaches for Plant I

Type	K_p	K_i	K_d	% OS	T_s
AFSA	2.5196	0.7260	3.7364	1.8	18s
PSO	2.3808	2.2871	11.6224	38	40s
ZN-PID	1.276	4.18	16.76	60	62s

TABLE 5.6
Performance Based on Different Optimization Approaches for Plant II

Type	K_p	K_i	K_d	% OS	T_s
AFSA	1.1064	1.10282	3.29	20	10s
PSO	1.2569	1.2680	3.5102	30	11s
ZN-PID	13.22	21.98	26.187	42	13s

TABLE 5.7
Performance Based on Different Optimization Approaches for Plant III

Type	K_p	K_i	K_d	% OS	T_s
AFSA	17.778	0.9254	15.9425	0	5s
PSO	8.0570	2.8915	4.7685	6	12s
ZN-PID	6.98	3.876	0.987	22	6s

REFERENCES

Astrom, K. J., & Hang, C. C. (1991). Toward intelligent PID control. *Automatica*, *28*(1), 1–9.

Azhar, Z. A. (2012). Internal plant manual, Universiti Teknologi PETRONAS, Malaysia, 2012.

Cai, Y. (2010). Artificial fish school algorithm applied in a combinatorial optimization problem. *Intelligent Systems and Applications*, *1*, 37–43.

Gazi, V., & Passino, K. V. (2011). *Swarm stability and optimization* (pp. 1–8). Chennai, India: Springer.

Jiang, M., Wang, Y., Pfletschinger, S., Lagunas, M. A., & Yuan, D. (2007). Advanced intelligent computing theories and applications. With aspects of contemporary intelligent computing techniques. In Huang, D. S., Heutte, L., and Loog, M. (Eds.), *Optimal multiuser detection with artificial fish swarm algorithm* (pp. 1084–1093). CCIS 2, ICIC2007, Berlin, Heidelberg: Springer-Verlag.

Li, X. L., Shao, Z. J., & Qian, J. X. (2002). An optimizing method based on autonomous animals: Fish-swarm algorithm. *Systems Engineering Theory & Practice*, *22*(11), 32–38.

Li, X. L., Xue, Y. C., Lu, F., & Tian, G. H. (2004). Parameter estimation method based on artificial fish school algorithm. *Journal of Shan Dong University (Engineering Science), 34*(3), 84–87.

Lobato, F. S., Souza, D. L., & Gedraite, R. (2012). A comparative study using bio-inspired optimization methods applied to controllers tuning. *Frontiers in advanced control systems.* Ginalber Luiz de Oliveira Serra, https://www.scribd.com/document/122522051/Frontiers-in-Advanced-Control-Systems.

Luo, Y., Wei, W., & Wang, S. X. (2010). Optimization of PID controller parameters based on an improved artificial fish swarm algorithm. *Proceedings of Third International Workshop on Advanced Computational Intelligence*, August 25–27, Suzhou, Jiangsu, China.

Luo, Y., Zhang, J., & Li, X. (2007). Optimization of PID controllers using artificial fish swarm algorithm. *Proceedings of International Conference on Automation and Logistics*, August 18–21, Jinan, China.

Reza, A. (2014). Empirical study of artificial fish swarm algorithm. *International Journal of Computing, Communications and Networking, 3*(1), 1–7.

Rocha, A. M. A. C., Martins, T. F. M. C., & Fernandes, E. M. G. P. (2011). An augmented Lagrangian fish swarm based method for global optimization. *Journal of Computational and Applied Mathematics, 235*, 4611–4620.

Shen, W., Guo, X., Wu, C., & Wu, D. (2011). Forecasting stock indices using radial basis function neural networks optimized by artificial fish swarm algorithm. *Knowledge-Based Systems, 24*, 378–385.

Tsai, H. C., & Lin, Y. H. (2011). Modification of the fish swarm algorithm with particle swarm optimization formulation and communication behavior. *Applied Soft Computing Journal, 11*, 5367–5374.

Visioli, A. (2006). *Practical PID control* (pp. 1–20). London: Springer.

Wang, C. R., Zhou, C. L., & Ma, J. W. (2005). An improved artificial fish-swarm algorithm and its application in feedforward neural networks. *Proceedings of the Fourth International Conference on Machine Learning and Cybernetics*, August 18–21, (pp. 2890–2894), Guangzhou, China.

Wang, X., Gao, N., Cai, S., & Huang, M. (2006). An artificial fish swarm algorithm based and ABC supported QoS unicast routing scheme in NGI. In Min, G. et al. (Eds.), *Lecture Notes in Computer Science, ISPA 2006* (Vol. 4331, pp. 205–214). Berlin, Heidelberg: Springer-Verlag.

Yin, Z. F., & Tian, Y. F. (2006). IIR digital filter design based on artificial fish swarm algorithm. *Information Technology, 7*, 78–81.

Zheng, G., & Lin, Z. (2012). A winner determination algorithm for combinatorial auctions based on hybrid artificial fish swarm algorithm. *Physics Procedia, 25*, 1666–1670.

Zhu, W., Jiang, J., Song, C., & Bao, L. (2011). Clustering algorithm based on fuzzy C-means and artificial fish swarm. *Procedia Engineering, 29*, 3307–3311.

6 Multiobjective Optimization of Bioethanol Pretreatment Process

Multiobjective (MO)-based situations are increasingly being encountered in process optimization applications. Hence, it is imperative that a decision maker in the plant environment has multiple solution selections to decide on the most optimal solution. In this chapter, a real-world industrial scale problem, that is, MO bioethanol pretreatment was solved using a number of algorithms such as genetic algorithm (GA), gravitational search algorithm (GSA), particle swarm optimization (PSO), and differential evolution (DE). In addition, ideas from chaos theory were introduced to enhance the DE algorithm to become chaotic DE (CDE). Performance metrics such as convergence, diversity, and hypervolume indicator (HVI) were used in the analysis of the dominance of the solution set produced by these algorithms. Studies on the performance as well as the solution quality produced by these algorithms are presented in this chapter.

6.1 INTRODUCTION

Over the past years, MO optimization techniques have been implemented in numerous engineering applications in the industry. MO optimization can be generally divided into two classes: "bi-objective" and "more than two objectives" problems. This chapter discusses the problems involving chemical processes which contain more than two objectives.

In Sankararao and Gupta (2007), the MO optimization of an industrial fluidized-bed catalytic cracking unit (FCCU) was carried out. In this work, a "jumping gene" adaptation was done on the multiobjective simulated annealing (MOSA) and thus creating a new algorithm: the jumping gene MOSA (or MOSA-jG). These adaptations were shown to improve the algorithmic performance and reduce computational time. The authors tested their new algorithm on three bi-objective test cases and then went on to solve two formulations of the industrial FCCU problem. The first formulation was a bi-objective problem while the second was a triple-objective problem subject to operational constraints. In Lim, Floquet, and Joulia, (2001), the normal boundary intersection (NBI) approach was hybridized with the summation of weighted objective functions (SWOF) method to solve an industrial MO optimization problem. The problem considered was a bi-objective chemical process optimization problem where the objectives were set to minimize cost and maximize pollution prevention. It was also stated that by using this hybrid NBI approach, reliable optimization results were obtained even in the nonconvex sectors of the Pareto frontier. Using this hybrid approach, an effective trade-off between the objectives was obtained by the authors from the Pareto frontier. Similarly, a MO optimization of a steam reformer performance using GA was done by Rajesh, Gupta, Rangaiah, and Ray (2000). Here, an adaptation was introduced to the non-dominated sorting genetic algorithm (NSGA) for MO optimization applications. In that work, the two objectives considered were the minimization of the methane feed rate and the maximization of the flow rate of carbon monoxide in the syngas stream while taking into account process requirements, heat integration, and economics (constraints). By this MO optimization procedure, the authors aimed to minimize the operating costs and maximize profits. The MO optimization procedure was done successfully and the Pareto optimal operating conditions were obtained.

A MO chemical process problem for the thermal processing of food was solved by Sendín, Alonso, and Banga (2010). In that work, two case studies (bi-objective and triple-objective problems) with nonlinear dynamic models were tackled. A hybrid approach consisting of the weighted Tchebycheff and the NBI approaches was utilized. The MO optimization using this novel approach was successful in constructing the Pareto optimal set for the thermal processing of foods. In the works of Martinsa and Costa (2010), a MO optimization of a benzene production process by toluene hydrodealkylation was done. In that work, the MOSA algorithm was utilized to solve this bi-objective problem. The objectives in that work were to maximize the net present value and minimize potential environmental impact of the process. By using the MOSA approach, the authors managed to construct a Pareto optimal set and identify the optimal solutions to the benzene production process. In Salari, Naiei, and Nabavi (2008), the MO optimization of an ethane thermal cracking reactor was carried out. The NSGA-II approach was used for MO optimization where a bi-objective problem involving ethylene conversion and selectivity was maximized. The Pareto frontier was obtained by this method and authors also discussed extensively on the effects of the decision variables on the objectives of this process. MO scenarios have also surfaced in the field of separation process. For instance, in Fiandaca and Fraga (2009), MOSA technique was used to optimize the design of the pressure swing adsorption process (cyclic separation process). The design problem considered in that work involved the bi-objective maximization of the nitrogen recovery and nitrogen purity. The results obtained in that work provided a good approximation of the Pareto frontier with acceptable trade-offs between the objectives.

In this chapter, the details on the model as well as the strategies employed to optimize the MO bioethanol pretreatment process are discussed. The solutions obtained using the GA, GSA, PSO, DE, and CDE were analyzed using the performance metrics such as convergence, diversity, and HVI.

6.2 BIOETHANOL PRETREATMENT

The pretreatment technique optimized in this chapter is based on the work by Banerjee et al. (2009). The existence of significant amounts of sugars such as holocellulose (57–61% by weight) in rice husks makes it a good potential source to be utilized for the production of bioethanol. Rice husk is a form of lignocellulosic biomass. Lignocelluloses are a complex structure made up of cellulose, hemicelluloses, and lignin. These complex structures are resistant to degradation and limit the capabilities of potential sources like rice husk for methanol production (Zaldivar, Nielsen, & Olsson, 2001). Therefore, complex structures like lignocellulose require a series of treatments to release their constituent monomer structured sugars. These sugars can then be fermented to produce ethanol. The reduction of the biomass particle size, maximizing lignin removal, limiting the formation of inhibitors (in the form of degradation products), minimizing the loss of pentose and hexoses, and reducing the overall cost of the process was achieved through the research work by Mosier et al. (2005).

The pretreatment technique optimized in the work of Banerjee et al. (2009) is the "wet air oxidation" technique. This technique consumes very low amounts of fuel and is low in terms of operation costs. This method is a potentially effective pretreatment technique for fractionating lignocellulose into a solubilized hemicellulose fraction and a solid cellulose-rich fraction with minimum inhibitor formation. This then encourages enhanced enzymatic hydrolysis of the material (prior treatment) for the subsequent ethanol fermentation while having minimal inhibitor formation (Schmidt & Thomsen, 1998; Klinke, Ahring, Schmidt, & Thomsen, 2002).

Thus, the MO optimization of the pretreatment process of rice husk using the wet air oxidation method to obtain a cellulose-rich substrate amenable to further enzymatic hydrolysis was the main theme of the work by Banerjee et al. (2009). In their work, the data from the experiments were used to build a MO optimization model through multiple regression analysis. A block diagram that depicts the wet air oxidation process with the manipulated variables as the inputs and the objectives as the outputs is shown in Figure 6.1.

This model and the associated constraints are generally presented as follows:

$$\begin{aligned}
&\text{Maximize} \rightarrow \text{Cellulose yield}, f_1\\
&\text{Maximize} \rightarrow \text{Lignin Removal}, f_2\\
&\text{Maximize} \rightarrow \text{Hemicelluloses Solubilization}, f_3\\
&\text{subject to process constraints.}
\end{aligned} \tag{6.1}$$

The objective functions (which are cellulose yield in [%], lignin removal in [%], and hemicelluloses solubilization in [%]) are as follows:

$$\begin{aligned}
f_1 = &-38.3467 + 0.6179X_1 + 1.7429X_2 + 3.0846X_3\\
&\quad 0.0177X_1X_2 \quad 0.0206X_1X_3 + 0.0937X_2X_3
\end{aligned} \tag{6.2}$$

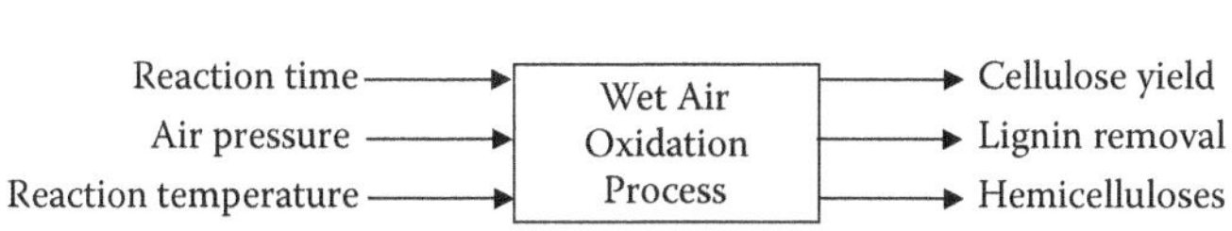

FIGURE 6.1 Block diagram of the wet air oxidation process.

$$f_2 = -3.8678 + 0.5587X_1 + 18.8545X_2 - 6.9167X_3 - 0.1066X_1X_2 + 0.0347X_1X_3 + 0.0377X_2X_3 \quad (6.3)$$

$$f_3 = -330.757 + 2.257X_1 - 9.612X_2 + 23.287X_3 + 0.033X_1X_2 - 0.134X_1X_3 + 0.26X_2X_3 \quad (6.4)$$

where X_1 is the reaction temperature (°C), X_2 is air pressure (MPa), and X_3 is the reaction time (minutes). The decision variables are constrained as per the experimental setup and the constraints are as follows:

$$X_1 \in [170, 195]$$
$$X_2 \in [0.5, 1] \quad (6.5)$$
$$X_3 \in [10, 20]$$

6.3 ALGORITHMS

In this work, the GA, PSO, GSA, and DE were employed to solve the MO oxidation process optimization problem. In addition, for rigorous analyses the hybrid PSO–DE approach was also implemented to the problem. The results achieved are discussed further in Section 6.5.

6.3.1 Genetic Algorithm

GAs are categorized as a class of population-based search and optimization algorithms. Detailed information about GA could be found in Holland (1992), Melanie (1996), Hugh (1993), David (1999a,b), and Lance (2000). The parameter settings initialized prior to the execution of the GA used in this work are shown in Table 6.1 and the GA program applied in this work is shown in Figure 6.2.

TABLE 6.1
Genetic Algorithm (GA) Settings

Parameters	Values
Length of individual string	6 bit
No. of individuals in the population	6
Probability of mutation	0.3333
Probability of recombination	0.5
Initial string of individuals	Random
Bit type of individual's string	Real-coded
Crossover type	N-point
Mutation type	N-bit flip
Selection type	Tournament

6.3.2 Particle Swarm Optimization and Gravitational Search

Since its introduction in 1995 by Kennedy and Eberhart (1995), the PSO algorithm could be found in many applications. It is essentially based on the concept of swarming (or flocking) behaviors of certain species of organisms (such as birds, fishes, etc.). Detailed description of the PSO algorithm could be found in Maurice (2006), Konstantinos and Michael (2010), Aleksandar (2009), Acar (2012), and Said and Ahmed (2008). The initialization parameters for the PSO algorithm are shown in Table 6.2 and the execution scheme is provided in Figure 6.3.

Inspired by the law of gravity and the mass interactions, the GSA algorithm was first developed in 2009 by Rashedi, Nezamabadi-pour, and Saryazdi (2009). This algorithm uses the Newtonian gravitational laws where the search agents are associated with massive objects. Thus, the gravitational forces influence the motion of these objects where lighter objects gravitate toward the heavier objects during these interactions. The gravitational force hence acts as the communication mechanism for the objects (Cui & Gao, 2011) similar to the "group component" for the particles in the PSO algorithm. The fitness landscape is closely linked to the position of the masses. During execution, the masses would

Step 1: Initialize a random chromosome for n individuals in the population

Step 2: Assign fitness conditions to each of the n individuals in the population

Step 3: By recombination from the current population, create offspring for the next generation

Step 4: Mutate offspring for this generation

Step 5: The parent selection to create the next generation is done by tournament selection

Step 6: The next population of n individuals is chosen

Step 7: Set new population to current population

Step 8: Assess the fitness of each offspring in the generation

Step 9: If the stopping criterion is satisfied halt program and print solutions, else go to Step 3

FIGURE 6.2 The GA program.

TABLE 6.2
PSO Settings

Parameters	Values
Initial parameter (c_1, c_2, r_1, r_2, w)	(1, 1.2, 0.5, 0.5, 0.8)
Number of particles	6
Initial social influence (s_1, s_2, s_3, s_4, s_5, s_6)	(1.1, 1.05, 1.033, 1.025, 1.02, 1.017)
Initial personal influence (p_1, p_2, p_3, p_4, p_5, p_6)	(3, 4, 5, 6, 7, 8)

gravitate to their fittest position reaching the optima. The initial parameters for the GSA algorithm are shown in Table 6.3 and the execution scheme is provided in Figure 6.4.

If during the iteration process, the maximum fitness has been reached such that the fitness criterion (no further optimization occurs in the objective function, no constraints are broken, and all the decision variables are nonnegative) is satisfied, the program is halted, and results are printed.

Step 1: Set number of particles, i and the initialize parameter settings w, c_1, c_2, r_1, r_2, n_o

Step 2: Randomly initialize particle's position $x_i(t)$ and velocity $v_i(t)$

Step 3: Calculate cognitive and social components of the particles

Step 4: Compute position $x_i(t+1)$ and velocity $v_i(t+1)$ of the particles at next iteration

Step 5: Proceed with the evaluation of the fitness of each particle in the swarm

Step 6: If the fitness conditions are satisfied and $t < T_{max}$, stop program and print solutions, else go to Step 3

FIGURE 6.3 The PSO program.

TABLE 6.3
GSA Parameter Settings

Parameters	Values
Initial parameter (G_0)	100
Number of mass agents, n	6
Constant parameter, α	20
Constant parameter, ε	0.01

Step 1: Initialize number of particles, algorithm parameters, and initial positions

Step 2: Compute gravitational & inertial masses and determine the gravitational constant

Step 3: Compute distance between agents

Step 4: Compute total force and the acceleration of each agent

Step 5: Compute new velocity and position for each agent

Step 6: Evaluate fitness condition for each of the agents (candidate solution)

Step 7: If the fitness criterion is satisfied and $t = T_{max}$, halt and print solutions, else proceed to step 3

FIGURE 6.4 The GSA program.

6.3.3 DE and Its Chaotic Enhancement

This pivotal idea of this technique is the incorporation of perturbative methods into evolutionary algorithms (Storn & Price, 1995). DE initializes a population of at least four real-coded vectors with some size *N* (individuals). The initial population of individual vectors is randomly generated in appropriate search ranges where one principal parent and three auxiliary parents are randomly selected from the population. Each individual in the population would have a chance to become a principal parent at one generation or the other. Thus, the principal parent has a chance of mating with the auxiliary parents engaged in "differential mutation" to generate a mutated vector. Using knock-out competition, the fitness evaluation scheme for the DE algorithm is carried out. The parameter setting for the DE algorithm is given in Table 6.4 and the DE execution scheme is shown in Figure 6.5.

The diversification mechanism used in this work is based on the concept of chaos theory (Lorenz, 1963). The idea here is to use chaos-based methods to increase the diversity of the population of solutions produced by the algorithm. In mathematical physics, chaotic systems are dynamical systems that are deterministic, behave in an irregular manner, are highly sensitive to initial conditions, and are

TABLE 6.4
Differential Evolution (DE) Parameter Settings

Parameters	Values
Individual size, *N*	6
Population size, *P*	7
Mutation amplification factor, *F*	0.3
Crossover probability, *CR*	0.667

Step 1: Set initial individual size and randomly initialize the population vectors

Step 2: Randomly select one principal parent and three auxilary parents

Step 3: Generate vector via differential mutation

Step 4: Recombine mutated vector with principal parent to generate child trial vector

Step 5: Perform "knock-out" competition for next generation survival selection

Step 6: If maximum number of iteration is reached and the fitness criterion is satisfied, stop program and print solutions,
else repeat step 2

FIGURE 6.5 The DE program.

impossible (or difficult) to predict in the long term (Flake, 1998). In this work, a one-dimensional chaotic map was used to increase the diversity of the population of solutions by embedding the map into the random number generation component in the algorithm. The one-dimensional chaotic map is represented as follows:

$$\psi_{n+1} = f(\psi_n) \tag{6.6}$$

The most widely studied one-dimensional map is the logistic map (Jakobson, 1981) which is as follows:

$$f(\psi_n) = r_n \psi_n (1 - \psi_n) \tag{6.7}$$

$$r_{n+1} = r_n + 0.01 \tag{6.8}$$

where $\psi_n \in [0,1]$ and $r \in [0,5]$. In this mapping, like all chaotic maps, the dynamics of the system varies for different sets of initial conditions (x_0 and r_0).

Augmentations were performed in the DE algorithm to enhance its diversification capabilities by the addition of the chaotic component. First, the population of vectors, x_i^G, was generated. The consequent steps are similar to the regular DE algorithm where one principal parent, x_i^p, and three auxiliary parents, x_i^a, are randomly selected. Differential mutation is then performed and the mutated vector, V_i, is generated. The V_i is then recombined with x_i^p to generate child trial vector, x_i^{child}. The obtained x_i^{child} is used as the input to the chaotic logistic map. This chaotic logistic mapping is presented as follows:

$$N_i(t) = x_i^{child}(t) \tag{6.9}$$

$$R_i(t) = \lambda N_i(t) \tag{6.10}$$

$$N_i(t+1) = R_i(t) N_i(t)[1 - N_i(t)] \tag{6.11}$$

$$R_i(t+1) = R_i(t) + \lambda' \tag{6.12}$$

where $N(t)$ and $R(t)$ are variables in the logistic chaotic map, λ and λ' are relaxation constants specified by the user. Then the logistic mapping is continued until a specific number of iterations is satisfied. The final value at maximum number of iteration of $N(t_{max})$ is incorporated into the child trial vector, x^{child}. Hence, the child trial vector, x_i^{child}, undergoes another round of mutation by the chaotic map. Next, the "knock-out" competition for next-generation survival selection is performed. The fitness function for the child trial vector, x_i^{child}, is evaluated. Thus, another variant of the DE algorithm with enhanced diversification capabilities is developed. In this work, this algorithm is called the CDE. The parameter settings specified in the CDE algorithm are shown in Table 6.5 and the execution scheme is given in Figure 6.6.

TABLE 6.5
Parameter Settings for the CDE Algorithm

Parameters	Values
Individual size, N	6
Population size, P	7
Mutation amplification factor, F	0.15
Maximum number of function evaluations, T_{max}	3000
Constant, λ	5
Constant, λ'	0.01
N_{max}	400

Step 1: Initialize population size

Step 2: Randomly initialize the population vectors, x_i^G

Step 3: Randomly select one principal parent, x_i^P

Step 4: Randomly select three auxilary parents, x_i^a

Step 5: Perform differential mutation & generate mutated vector, V_i

Step 6: Recombine V_i with x_i^P to generate child trial vector, x_i^{child}

Step 7: Iterate chaotic logistic map

Step 8: If $n > N_{max}$, proceed to next step,
else go to step 7

Step 9: Evaluate fitness of the new x_i^{child}

Step 10: If the fitness criterion is satisfied halt and print solutions,
else proceed to step 3

FIGURE 6.6 The CDE program.

6.4 SOLUTION MEASUREMENT

In this work, the performance metrics comprise of convergence, diversity, and HVI that will be discussed briefly in the following sections. The convergence metric used in this work was developed in Deb and Jain (2002). This metric gauges the convergence property of a solution set with respect to a reference set. For this convergence metric, low metric values indicate high convergence characteristics among the solution vectors. The diversity metric used in this work is the sigma diversity metric (SDM) (Mostaghim & Teich, 2003). The SDM evaluates the locations of the solution vectors in the objective space relative to the sigma vectors. The high values of the SDM indicate high uniformity and diversity in terms of the distributions of the solution vectors in the objective space. One approach that has been effective in measuring the quality of the solution set that constructs the Pareto frontier in cases where the Pareto frontier is unknown is the HVI. The HVI is the only indicator which is strictly Pareto compliant that can be used to measure the quality of solution sets in MO optimization problems. Further details about this can be found in Beume, Naujoks, and Emmerich (2007) and Zitzler and Thiele (1999).

6.5 COMPUTATIONAL RESULTS

First, the solution characteristics in the objective space are attained. This is done by determining the convergence and diversity values of all the algorithms when applied to this problem. The values given by the convergence metric for all the algorithms in this study are as shown in Figure 6.7 and Table 6.6.

As mentioned previously, the smaller the value of the convergence metric, the more convergent the solutions are on Pareto frontier. From Figure 6.7 and Table 6.6, it can be observed that the Pareto frontier produced by the GSA algorithm is the most convergent followed by the DE algorithm and the GA algorithm, respectively. The least convergent frontier was produced by the PSO algorithm.

The values given by the SDM for all the mentioned algorithms are shown in Figure 6.8 and Table 6.7.

As for the diversity metric, the higher its value, the more diverse the solutions are on Pareto frontier. From Figure 6.8 and Table 6.7, it can be observed that the Pareto frontier produced by the PSO algorithm is the most diverse followed by the GSA algorithm. The DE and the GA algorithms are equivalent in this respect. The values given by the HVI for all the mentioned algorithms are shown in Figure 6.9 and Table 6.8.

The HVI functions such that the higher its value, the more dominant the Pareto frontier. From Figure 6.9 and Table 6.8, it can be observed that the Pareto frontier

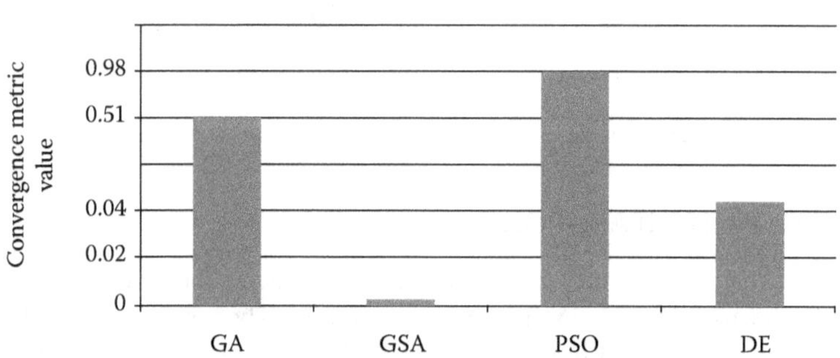

FIGURE 6.7 Values of the convergence metric for each algorithm.

TABLE 6.6
Values of the Convergence Metric for Each Algorithm

GA	GSA	PSO	DE
0.519	0.003	0.987	0.045

produced by the DE algorithm gives the most dominant frontier followed by the PSO algorithm and the GSA algorithm, respectively. The least dominant frontier was produced by the GA algorithm.

Due to the application of the NBI method, multiple runs were executed for each algorithm to obtain the individual optima. Then, the β-subproblem was solved on a

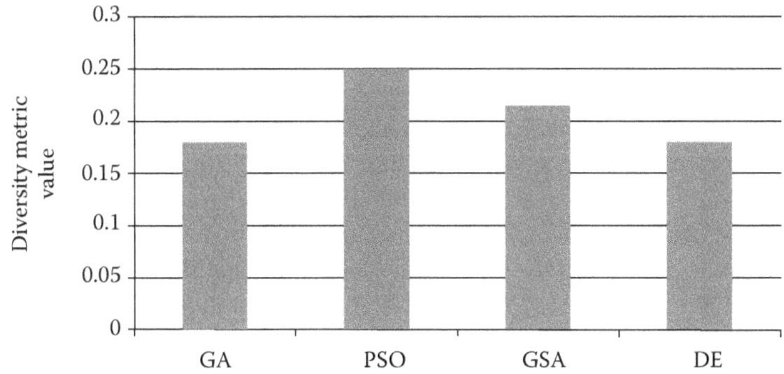

FIGURE 6.8 Values of the sigma diversity metric for each algorithm.

TABLE 6.7
Values of the Sigma Diversity Metric for Each Algorithm

GA	PSO	GSA	DE
0.179	0.25	0.214	0.179

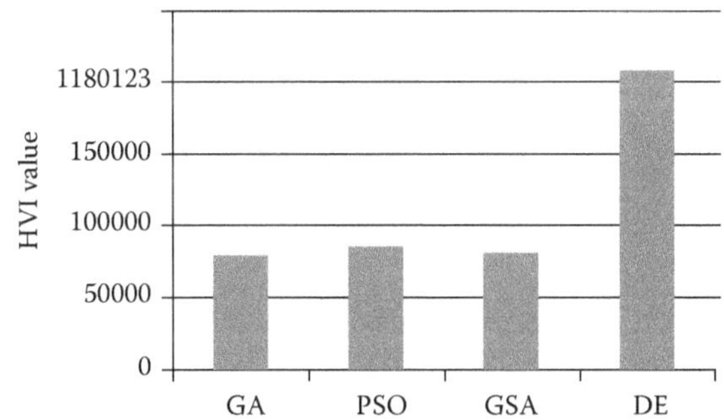

FIGURE 6.9 Values of the HVI for each algorithm.

TABLE 6.8
Values of the HVI for Each Algorithm

GA	PSO	GSA	DE
79,999.09	86,033.32	81,697.74	208,012

single run by each algorithm. To calculate the total computational time, the average computational time for the multiple runs to obtain the individual optima is summed with the computational time taken to solve the β-subproblem. The lowest computational time during execution was achieved by GSA (5.104 s), followed by the GA (6.391 s) and the DE (13.742 s) algorithms. The PSO algorithm (20.926 s) took the longest computational time during execution. The Pareto frontier produced by the GA, GSA, DE, and PSO algorithms when applied to this application is presented graphically in Figure 6.10.

The next step is to analyze the influence of diversity and convergence of the solutions on the Pareto frontiers (produced by each algorithm) on the degree of dominance. From the HVI values, two algorithms with the lowest values can be eliminated from further analysis since the Pareto frontiers produced by these algorithms are the least dominant. In this case, the GSA and the GA algorithms are these candidates (see Figure 6.10). It can be observed that the DE algorithm is the best and the PSO algorithm is the second best algorithm in terms of Pareto frontier dominance.

In any optimization problem, an optimal solution (or for MO scenarios, the most dominant Pareto frontier) can only be achieved if the degree of diversity and convergence of the solution sets produced by an algorithm matches the characteristics of the objective space. Keeping this in mind, the degree of diversity and convergence of the solution sets that construct the Pareto frontier influences the degree of dominance measured by the HVI metric.

In Figure 6.10, it can be observed that the PSO algorithm produces solutions that are less convergent but more diverse than the DE algorithm. Enhancement is done on the algorithm with the highest HVI (which in this case is the DE algorithm) to boost

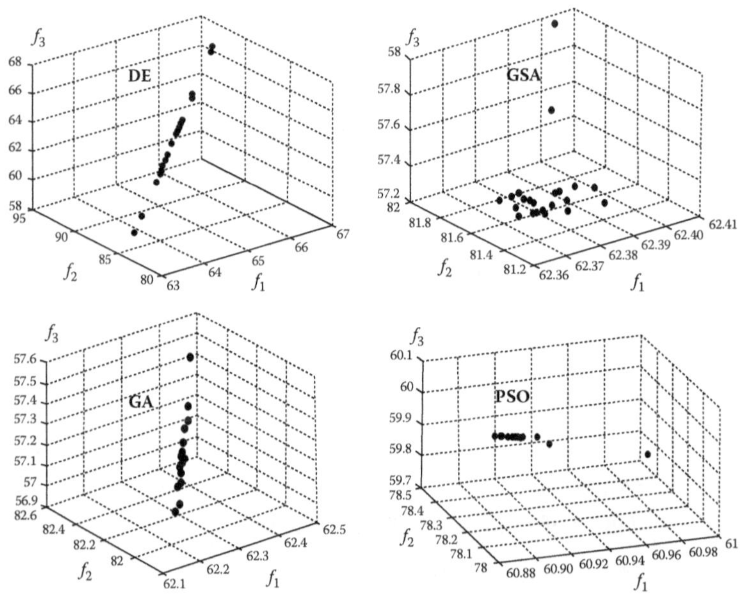

FIGURE 6.10 Pareto frontiers for the GA, GSA, DE, and PSO algorithms.

further its performance. From observing the second best algorithm (PSO) in terms of dominance (HVI), necessary enhancements to the DE algorithm are carried out. By observing the information of the solution characteristics given by the solution sets produced by the DE algorithm, appropriate modifications can be carried out. It can be seen that the DE algorithm produces the solution set which is more convergent than the PSO algorithm. Hence, in this respect, no modifications can be done to the DE algorithm. However, it can also be observed that the solution set produced by the PSO algorithm is more diverse than the DE algorithm. Hence, the DE algorithm lacks in terms of diversity of the solution set. Therefore, the DE algorithm is modified such that the diversity of its solutions increases. This way, the DE algorithm is modified to match the diversity properties of the PSO algorithm.

Thus, an additional component is incorporated into the DE algorithm that increases its capabilities to produce solutions that are more diverse. By improving its capabilities to produce diverse solutions, a higher degree of dominance of the Pareto frontier would be achieved. By this interpretation, the diversity properties of the most dominant algorithm (in this case: DE) is improved by adding the chaotic component to the original DE to obtain the CDE algorithm. The values given by the convergence metric, diversity metric, and the HVI for the DE and CDE algorithms are shown in Figures 6.11 through 6.13 respectively:

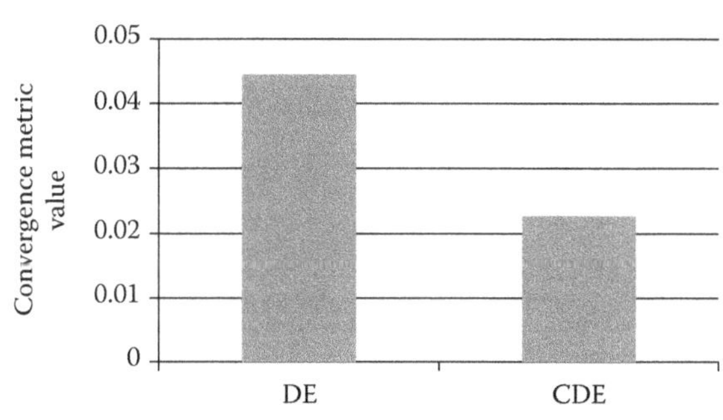

FIGURE 6.11 Values of the convergence metric for DE and CDE.

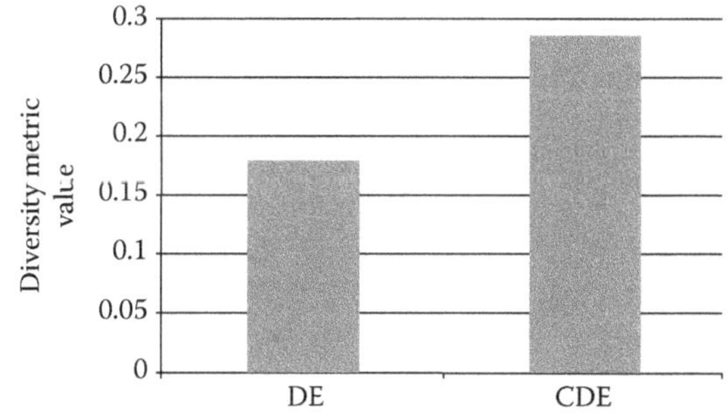

FIGURE 6.12 Values of the diversity metric for DE and CDE.

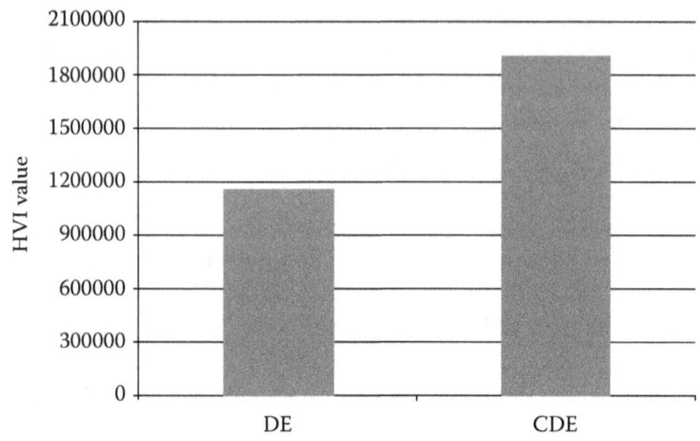

FIGURE 6.13 Values of the HVI for DE and CDE.

TABLE 6.9
Values of the HVI, Convergence, and Diversity Metrics for DE, CDE, and PSO–DE Algorithms

	HVI	Convergence Metric Value	Diversity Metric Value
DE	1,180,123	0.045	0.179
CDE	1,922,933	0.022	0.286
PSO–DE	82,063.8	0.017	0.107

To provide more rigorous analyses, the hybrid PSO–DE technique was applied to the MO application problem. Some of the results presented below involve the PSO–DE approach. The metric values given by the HVI for the DE, CDE, and PSO–DE algorithms are shown in Table 6.9.

It can be observed from Figure 6.13 and Table 6.9 that the diversity level of the Pareto frontier produced by the CDE algorithm is higher than that by the DE algorithm. In terms of convergence (see Figure 6.11), the Pareto frontier produced by the DE algorithm is better than the CDE algorithm. As can be observed in Figure 6.12 and Table 6.9, the HVI value of the Pareto frontier produced by the CDE algorithm is more dominant than the DE algorithm by 62.944%. Hence, it can be inferred that the solution diversity at the Pareto frontier heavily affects the degree of dominance.

The Pareto frontier produced by the CDE and the PSO–DE algorithms is presented graphically in Figures 6.14 and 6.15 respectively.

The computational time for the execution of the CDE and PSO–DE algorithms are 33.48 and 66.301 s, respectively. Hence, the CDE algorithm is more computationally efficient as compared to the PSO–DE algorithm since the PSO–DE algorithm takes the most computational time. The comparison of the individual best solution options obtained by the DE, CDE, and the MINITAB 14 method (used in Banerjee et al., 2009) is shown in Table 6.10.

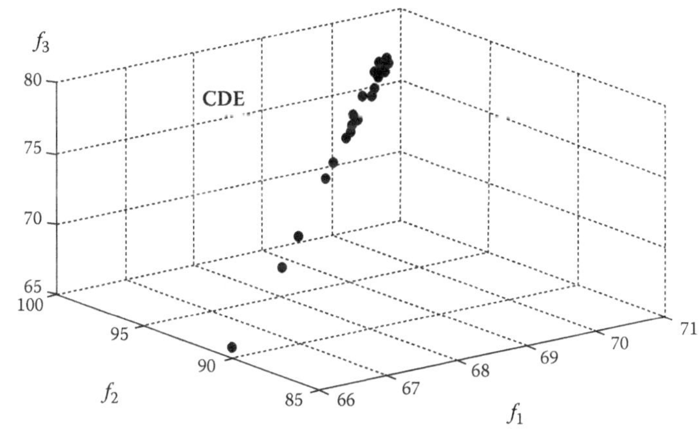

FIGURE 6.14 The Pareto frontiers of the CDE algorithm.

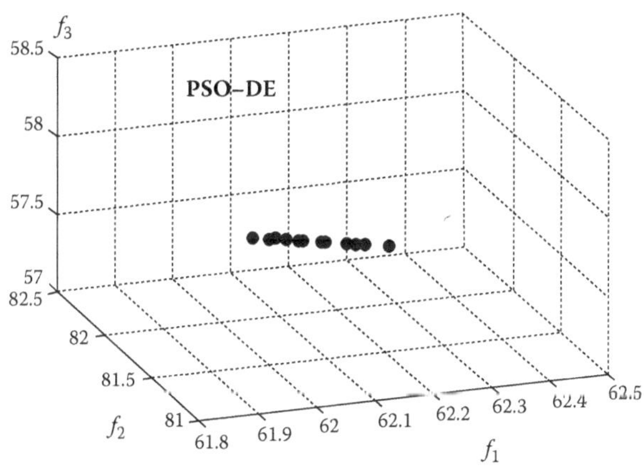

FIGURE 6.15 The Pareto frontiers of the PSO–DE algorithm.

TABLE 6.10
Comparison of Individual Best Solution Options

Objective Functions, Constraints, and Performance Metric	DE	CDE	Banerjee et al. (2009)
f_1	66.414	70.753	66.97
f_2	91.083	99.793	89
f_3	67.211	76.872	69.77
x_1	181.976	191.93	185
x_2	0.999	0.756	0.5
x_3	10.878	10.326	15
HVI	56,792.98	79,436.63	13,084.38

It can be observed in Table 6.10 that the individual solution produced by the CDE and the DE methods employed in this work outweighs the solution produced by the method used in Banerjee et al. (2009) by 507.11% and 334.057%, respectively. In this industrial application, it can be observed that the CDE method outperforms all other methods in terms of Pareto frontier dominance. It can be stated that the CDE method is more computationally efficient and gives better solution quality as compared to the hybrid approach, the PSO–DE method. All algorithms applied in this application performed stable computations during the search of the individual optima as well as while solving the β-subproblems (aggregated objective function). All Pareto-efficient solutions produced by the algorithms developed in this work were feasible and no constraints were compromised. The advantage of using the CDE algorithm as compared to the other algorithms in this application is that it produces solutions that are highly dominant in terms of MO optimization of the process parameters.

6.6 SUMMARY

In this chapter, six algorithms (GA, PSO, GSA, DE, PSO–DE, and CDE) were employed in optimization of MO problem of bioethanol pretreatment process. The parameter settings as well as the details regarding algorithms and executions were described. In addition, three measurement metrics such as convergence metric, diversity metric, and HVI were used to gauge the performance of the algorithms. It was found that the CDE outperformed all the other algorithms.

REFERENCES

Acar, V. R. (2012). *Particle swarm optimization for electronic circuit design automation.* PAV Akademikerverlag GmbH & Co. KG, https://www.morebooks.de/store/gb/book/particle-swarm-optimization-for-electronic-circuit-design-automation/isbn/978-3-659-25268-6.

Aleksandar, L. (2009). *Particle swarm optimization.* InTech, http://www.intechopen.com/books/particle_swarm_optimization.

Banerjee, S., Sen, R., Pandey, R. A., Chakrabarti, T., Satpute, D, Giri, B. S., et al. (2009). Evaluation of wet air oxidation as a pre-treatment strategy for bioethanol production from rice husk and process optimization. *Biomass & Bioenergy, 33*, 1680–1686.

Beume, N., Naujoks, B., & Emmerich, M. (2007). SMS-EMOA: Multiobjective selection based on dominated hypervolume. *European Journal of Operational Research, 181*(3), 1653–1669.

Clerc, M. (2006). *Particle swarm optimization.* London: Wiley-ISTE.

Cui, Z., & Gao, X. (2011). Theory and applications of swarm intelligence. *Neural Computing and Applications, 21*(2), 205–206.

David, A. C. (1999a). *An introduction to genetic algorithms for scientists and engineers.* Singapore: World Scientific Publishing.

David, G. (1999b). *Algorithms in search, optimization, and machine learning.* Reading, MA: Addison-Wesley.

Deb, K., & Jain, S. (2002). Running performance metrics for evolutionary multiobjective optimization, KanGAL Report No. 2002004. Kanpur Genetic Algorithms Laboratory, Indian Institute of Technology, Kanpur, India.

Fiandaca, G., & Fraga, E. S. (2009). A multi-objective genetic algorithm for the design of pressure swing adsorption. *Engineering Optimization*, *41*(9), 833–854.

Flake, G.W. (1998). *The computational beauty of nature: Computer explorations of fractals, chaos, complex systems, and adaptation* (pp. 469–482). Cambridge, MA: MIT Press.

Holland, J. H. (1992). *Adaptation in natural and artificial systems: An introductory analysis with applications to biology, control and artificial intelligence*. Cambridge, MA: MIT Press.

Hugh, M. C. (1993). *Applications of artifical intelligence in chemistry*. Oxford: Oxford University Press.

Jakobson, M. (1981). Absolutely continuous invariant measures for one-parameter families of one-dimensional maps. *Communications on Mathematical Physics*, *81*, 39–88.

Kennedy, J., & Eberhart, R. (1995). Particle swarm optimization, *IEEE Proceedings of the International Conference on Neural Networks* (pp. 1942–1948), Perth, Australia.

Klinke, H. B., Ahring, B. K., Schmidt, A. S., & Thomsen, A. B. (2002). Characterization of degradation products from alkaline wet oxidation of wheat straw. *Bioresource Technology*, *82*, 15–26.

Konstantinos, E. P., & Michael, N. V. (2010). *Particle swarm optimization and intelligence: Advances and applications*. IGI Global, http://www.igi-global.com/chapter/particle-swarm-optimization/40629.

Lance, C. (Ed.). (2000). *The practical handbook of genetic algorithms*. Boca Raton, FL: CRC Press.

Lim, Y. I., Floquet, P., & Joulia, X. (2001). Efficient implementation of the normal boundary intersection (NBI) method on multiobjective optimization problems. *Industrial Engineering and Chemical Research*, *40*, 648–655.

Lorenz, E. N. (1963). Deterministic non-periodic flow. *Journal of the Atmospheric Sciences*, *20*(2), 130–141.

Martinsa, F., & Costa, C. A. V. (2010). Multiobjective optimization with economic and environmental objective functions using modified simulated annealing. *20th European Symposium on Computer Aided Process Engineering—ESCAPE20*, *28*, 1–6, http://www.igi-global.com/chapter/particle-swarm-optimization/40629.

Melanie, M. (1996). *Introduction to genetic algorithms*. Cambridge, MA: MIT Press.

Mosier, N., Wyman, C., Dale, B., Elander, E., Lee, Y. Y., & Holtzapple, H. (2005). Features of promising technologies for pretreatment of lignocellulosic biomass. *Bioresource Technology*, *96*, 673–686.

Mostaghim, S., & Teich, J. (2003). Strategies for finding good local guides in multiobjective particle swarm optimization. In *2003 IEEE Swarm Intelligence Symposium Proceedings, IEEE Service Center, Indianapolis, USA* (pp. 26–33).

Rajesh, J. K., Gupta, S. K., Rangaiah, G. P., & Ray, A. K. (2000). Multiobjective optimization of steam reformer performance using genetic algorithm. *Industrial and Engineering Chemistry Research*, *39*, 706–717.

Rashedi, E., Nezamabadi-pour, H., & Saryazdi, S. (2009). GSA: A gravitational search algorithm. *Information Sciences*, *179*, 2232–2248.

Said, M. M., & Ahmed, A. K. (2008). *Particle swarm optimization—A physics-based approach*. Morgan and Claypool, http://www.morganclaypool.com/feedback/show.

Salari, D., Naiei, A., & Nabavi, S. R. (2008). Multi-objective genetic optimization of ethane thermal cracking reactor. *Iranian Journal of Chemical Engineering*, *5*(3), 29–39.

Sankararao, B., & Gupta, S. K. (2007). Multi-objective optimization of an industrial fluidized-bed catalytic cracking unit (FCCU) using two jumping gene adaptations of simulated annealing. *Computers and Chemical Engineering*, *31*, 1496–1515.

Schmidt, A. S., & Thomsen, A. B. (1998). Optimization of wet oxidation pretreatment of wheat straw. *Bioresource Technology*, *64*, 139–151.

Sendín, J. O. H., Alonso, A. A., & Banga, J. R. (2010). Efficient and robust multi-objective optimization of food processing: A novel approach with application to thermal sterilization. *Journal of Food Engineering*, *98*, 317–324.

Storn, R., & Price, K. V. (1995). Differential evolution—A simple and efficient adaptive scheme for global optimization over continuous spaces, ICSI, Technical Report TR-95-012.

Zaldivar, J., Nielsen, J., & Olsson, L. (2001). Fuel ethanol production from lignocellulose: A challenge for metabolic engineering and process integration. *Applied Microbiology & Biotechnology*, *56*, 17–34.

Zitzler, E., & Thiele, L. (1999). Multiobjective evolutionary algorithms: A comparative case study and the strength Pareto approach. *IEEE Transactions on Evolutionary Computation*, *3*(4), 257–271.

Part III

Material Engineering

7 Bioactive Compound Extraction Process Optimization

Multiobjective (MO) optimization is an emerging field which is increasingly being implemented in many process-related industries on a global scale. A near-complete understanding of the concerned process mechanisms is crucial for the formulation of the optimization model. Identifying the process-related uncertainties and the process characteristics (such as the existence of nonlinearities, variations in the input, and the process bounds) helps to strengthen the formulation of the process model. The MO models of extraction processes usually form a characteristically complex optimization problem which is relevant to many chemical and pharmaceutical industries. Finding an effective strategy for obtaining the optimal settings for the extraction process is central to solving such problems.

7.1 INTRODUCTION

Multicriteria or MO scenarios have become increasingly prevalent in industrial engineering environments (Statnikov & Matusov, 1995). MO optimization problems are commonly tackled using the concept of Pareto-optimality to trace out the non-dominated solution options at the Pareto curve (Zitzler & Thiele, 1998; Deb, Pratap, Agarwal, & Meyarivan, 2002). Other methods include the weighted techniques which involve objective function aggregation resulting in a master weighted function. This master weighted function is then solved for various weight values (which are usually fractional) (Fishburn, 1967; Luyben & Floudas, 1994; Das & Dennis, 1998; Triantaphyllou, 2000). Using these techniques, the weights are used to consign relative importance or priority to the objectives in the master aggregate function. Hence, alternative near-optimal solution options are generated for various values of the scalars. In this chapter, the normal boundary intersection (NBI) scheme (Das & Dennis, 1998) is presented as a scalarization tool to construct the Pareto frontier. In Sandgren (1994) and Statnikov and Matusov (1995), detail examples and analyses on MO techniques for problems in engineering optimization are presented.

Over the past years, MO optimization has been introduced and applied into many engineering problems. Some of these developments are briefly discussed in the following. In Aguirre, Zebulum, and Coello (2004), a MO evolutionary algorithm with an enhanced constraint handling mechanism was used to optimize the circuit design of a field programmable transistor array (FPTA). The authors used the inverted shrinkable Pareto archived evolution strategy (ISPAES) for the MO optimization of the circuit design. Another MO problem involving engineering design was solved by Reddy and

Kumar (2007). In that work, a MO swarm intelligence algorithm was developed by incorporating the Pareto dominance relation into the standard particle swarm optimization (PSO) algorithm. Three engineering design problems—the "two bar truss design" (Palli, Azram, McCluskey, & Sundararajan, 1999), "I-beam design" (Yang, Yeun, & Ruy, 2002), and the "welded beam design" (Deb, Pratap, & Moitra, 2000) problems—were successfully solved by Reddy and Kumar (2007). In the area of thermal system design, the MO optimization of an HVAC (heating, ventilating, air-conditioning, and cooling) system was carried out by Kusiak, Xu, and Tang (2010). In that work, a neural network was used to derive the MO optimization model. This model was then optimized using a multiobjective PSO algorithm (MOPSO). Using this algorithm, the authors identified the optimum control settings for the supply air temperature and static pressure to minimize the air handling unit energy consumption while maintaining air quality. Another application of the non-dominated sorting genetic algorithm (NSGA-II) to engineering system design was done by Nain, Giri, Sharma, and Deb (2010). In Nain et al. (2010), the authors optimized the structural parameters (area and length of the thermoelectric cooler [TEC] elements) of the TEC. The coefficient of performance (COP) and the rate of refrigeration (ROR) were successfully maximized in that work.

Recently, MO optimization methods have also expanded the power and energy industries. For instance, in Van Sickel, Venkatesh, and Lee (2008), the MO optimization of a fossil fuel power plant was done using multiobjective evolutionary programming (MOEP) and MOPSO algorithms. The MO techniques in that work were applied to develop reference governors for power plant control systems. MO optimization of reference governor design for power plants was done by Heo and Lee (2006). In that work, PSO variant algorithms were used to find the optimal mapping between unit load demands and pressure set point of a fossil fuel power plant. By this approach, the optimal set points of the controllers under a large variety of operation scenarios were achieved. Similarly, in the works of Song and Kusiak (2010), temporal processes in power plants were optimized using MO techniques. In that work, the central theme was to maximize the boiler efficiency while minimizing the limestone consumption. Two approaches—the data mining (DM) and evolutionary strategy algorithms—were combined to solve the optimization model. In Song and Kusiak (2010), the MO optimization of temporal-dependent processes were successfully completed by identifying the optimum control parameters. One other area in which MO optimization has been applied with considerable success is the field of economic/environmental dispatch for power systems. For instance, in the works of Gunda and Acharjee (2011), a MO economic/environmental dispatch problem was solved using the Pareto frontier differential evolution (PFDE) approach. By using this technique, the authors managed to minimize the fuel consumption and emissions with minimal energy loss. This triple-objective problem was successfully solved without the violation of the system's security constraints. Another similar problem was tackled in King, Rughooputh, and Deb (2005). In that work, power generation optimization was done to minimize the total fuel costs as well as the amount of emission.

The MO problem considered in this chapter was formulated by Shashi, Deep, and Katiyar (2010). This problem involves the optimization of the yields of certain chemical products which are extracted from the *Gardenia jasminoides Ellis* fruit. The MO optimization model was developed by Shashi et al. (2010) to maximize the extraction yields of the three bioactive compounds: crocin, geniposide, and total phenolic compounds. The optimal extraction parameters which construct the most dominant Pareto frontier are then identified such that the process constraints remain unviolated. In Shashi et al. (2010), the MO problem was tackled using the real-coded genetic algorithm (RCGA) approach to obtain a single individual optima and not a Pareto frontier. In that work, measurement metrics were not employed to evaluate the solution quality in detail. In addition, the work done in Shashi et al. (2010) focused on modeling the system rather than optimizing it. The authors of that work employed only one optimization technique and did not carry out extensive comparative analysis on the optimization capabilities. Due to the mentioned setbacks, these factors are systematically addressed in this chapter to provide some insights on the optimization of the extraction process.

Over the past years, swarm intelligence-based metaheuristic techniques have been applied with increasing frequency to industrial MO scenarios. Some of the most effective swarm approaches have been devised using ideas from Newtonian gravitational theory (Rashedi, Nezamabadi-pour, & Saryazdi, 2009), dynamics of fish movement (Neshat, Sepidnam, Sargolzaei, & Toosi, 2014), and bird flocking behaviors (Kennedy & Eberhart, 1995). In this chapter, two swarm-based techniques—PSO (Kennedy & Eberhart, 1995) and the novel Hopfield-enhanced PSO (HoPSO)—are presented and implemented to the extraction problem (Shashi et al., 2010). The measurement techniques such as the convergence metric (Deb & Jain, 2002) and the hypervolume indicator (HVI) (Zitzler & Thiele, 1998) were used to analyze the solution spread produced by these algorithms. The HVI is a set measure reflecting the volume enclosed by a Pareto front approximation and a reference set (Emmerich, Beume, & Naujoks, 2005). The convergence metric on the other hand measures the degree at which the solutions conglomerate toward optimal regions of the objective space. Using the values obtained by the measurement metrics, the correlation between the convergence and the degree of dominance (measured by the HVI) of the solution sets is obtained and discussed. The solutions constructing the Pareto frontier obtained using the developed HoPSO algorithm is also subjected to the analyses mentioned above.

In this chapter, the details on the model as well as the strategies employed to optimize the MO extraction process are discussed. In addition, the procedures involving the overall optimization basis called the NBI are explored. By the implementation of the measurement metrics, the solutions obtained using the metaheuristics are then gauged and analyzed.

7.2 MODEL REPRESENTATION

The model for MO problem considered in this chapter was developed by Shashi et al. (2010). This problem involves the optimization of the yields of certain chemical products which are extracted from the *G. jasminoides Ellis* fruit. The phenolic compounds in *G. jasminoides Ellis* have high antioxidant capabilities which make this fruit valuable for medicinal uses (Li, Wong, Chen, & Chen, 2008). Compared to other natural food pigments, the coloring constituents of the fruit of Gardenia are nontoxic and chemically stable (Van Calsteren et al., 1997). The constituents present in the Gardenia fruit (Oshima et al., 1988) are iridoid glycosides (e.g., gardenoside, geniposide, gardoside, and scandoside methyl ester). These constituents could be converted into blue colorants under aerobic conditions by enzymes or some microorganisms. The Gardenia fruit extract in its rudimentary form also contains phenolic compounds with high antioxidant capacity in abundance (Li et al., 2008). The MO optimization model in Shashi et al. (2010) was for the extraction process of bioactive compounds from Gardenia with respect to the constraints. The MO optimization model was developed to maximize the yield of three bioactive compounds: crocin (f_1), geniposide (f_2), and total phenolic compounds (f_3). This model is presented as follows:

$$\begin{aligned}
&\text{Maximize} \rightarrow \text{Crocin}, f_1 \\
&\text{Maximize} \rightarrow \text{Geniposide}, f_2 \\
&\text{Maximize} \rightarrow \text{Total phenolic compounds}, f_3 \\
&\text{subject to } \textit{process constraints}.
\end{aligned} \tag{7.1}$$

The objective functions represent the yields of each of the bioactive compounds in the units of mg/g of dry powder as in Shashi et al. (2010). The objective function modeled with respect to the decision variables are as follows:

$$\begin{aligned}
f_1 = {} & 3.8384907903 + 0.0679672610X_1 + 0.0217802311X_2 + 0.0376755412X_3 \\
& - 0.0012103181X_1^2 + 0.0000953785X_2^2 - 0.0002819634X_3^2 \\
& + 0.0005496524X_1X_2 - 0.0009032316X_2X_3 + 0.0008033811X_1X_3
\end{aligned} \tag{7.2}$$

$$\begin{aligned}
f_2 = {} & 46.6564201287 + 0.6726057655X_1 + 0.4208752507X_2 \\
& + 0.9999909858X_3 - 0.0161053654X_1^2 - 0.0034210643X_2^2 \\
& - 0.0116458859X_3^2 + 0.0122000907X_1X_2 - 0.0095644212X_2X_3 \\
& + 0.0089464814X_1X_3
\end{aligned} \tag{7.3}$$

$$\begin{aligned}
f_3 = {} & -6.3629169281 + 0.4060552042X_1 + 0.3277005337X_2 \\
& + 0.3411029105X_3 - 0.0053585731X_1^2 - 0.0020487593X_2^2 \\
& - 0.0042291040X_3^2 + 0.0017226318X_1X_2 - 0.0011990977X_2X_3 \\
& + 0.0007814998X_1X_3
\end{aligned} \tag{7.4}$$

The decision variables (which are the process extraction parameters) are constrained as per the experimental setup described in Shashi et al. (2010). The constraints are as follows:

$$\begin{aligned}
X_1 &\in [19.5, 80.5] \\
X_2 &\in [27.1, 72.9] \\
X_3 &\in [7.1, 52.9]
\end{aligned} \tag{7.5}$$

where X_1 is the concentration of ethanol in %, X_2 is the extraction temperature in °C, and X_3 is the extraction time in minutes.

7.3 METAHEURISTIC FRAMEWORK

7.3.1 Genetic Algorithms

The computational strategy employed for tackling this problem arises from two schools of thought which are swarm and evolutionary frameworks. Evolutionary intelligence originates from the idea presented in Holland's genetic algorithm. The central theme of all evolutionary algorithms is survival of the fittest (natural selection) which acts on a population of organisms under environmental stress. Thus, a fitness distribution is formed in the mentioned population of organisms (individuals). In the case of optimization, let there be an objective function to be optimized. Then, various random candidate solutions (individuals) to the problem can be gauged by applying a fitness function to each of these solutions (where the higher the fitness function, the better the solution quality with respect to the objective function). Each evaluated individual with respect to the best fitness values is then chosen for producing the next generation of potential solution vectors through the processes of crossover and mutation. The crossover operator crosses the parent individuals from the previous generation (fittest individuals) and produces the next generation of individuals. The mutation operator, on the other hand, perturbs the gene pool of the population to generate new optimization capabilities in the new individual offspring. Thus, all individuals compete in their population as they achieve higher fitness values. The repeated execution of this cycle is metaheuristic since the solution vectors in the population are subsequently improved as the iteration proceeds.

Genetic algorithm (GA) is one of the first evolutionary search and optimization techniques. This population-based approach uses an N-point crossover operator to create new offspring for successive generations. To avoid algorithmic stagnation at some local minima, mutation operators are usually employed to diversify the search. A bit flip-type mutation operator was utilized in the GA and the algorithmic scheme applied is provided in Algorithm 7.1. The flow of the GA is shown in Algorithm 7.1:

Algorithm 7.1: Genetic Algorithm

Step 1: Randomly initialize a population of n individuals
Step 2: Fitness criteria are assigned to each of the individuals
Step 3: Generate offspring by recombination from the current population
Step 4: Mutate offspring
Step 5: Perform parent selection (tournament selection)
Step 6: A new population of n individuals is selected
Step 7: Set new population = current population
Step 8: Evaluate fitness of offspring
Step 9: If the termination conditions are satisfied halt and print solutions, else go to step 3.

7.3.2 Particle Swarm Optimization

The PSO algorithm introduced in 1995 (Kennedy & Eberhart, 1995) springs from two distinct frames of ideas. The first concept was based on the examination of swarming (or flocking) behaviors of certain species of organisms (such as birds, fishes, etc.). The second idea sprung from the study of evolutionary computations. The PSO algorithm searches the search space for candidate solutions and evaluates these solutions with respect to some (user specified) fitness condition. The candidate optimal solutions obtained by this algorithm are achieved as a result of particles which are in motion (swarming) through the fitness landscape. In the beginning, some candidate solutions are selected by the PSO algorithm. These solutions can be randomly selected or be established with the aid of some a priori facts. Next, the evaluation of the particles' position and velocity (which are also the candidate solutions) relative to the fitness function is carried out. Consequently, in conjunction with the fitness function a condition is introduced, where, if the fitness function is not fulfilled, then the algorithm updates the individual and social terms by the aid of a user-specified update rule. Following this, the velocity and the position of the particles are updated. This recursive course of action is iterated until the fitness function is satisfied by all candidate solutions and solutions have thus converged into a fixed position. It is essential to note that the velocity- and position-updating rule is critical to the optimization capabilities of this method. The velocity of each particle in motion (swarming) is updated using the following equation:

$$v_i(t+1) = wv_i(t) + c_1r_1\left[\hat{x}_i(t) - x_i(t)\right] + c_2r_2\left[g(t) - x_i(t)\right] \tag{7.6}$$

where each particle is identified by the index i, $v_i(t)$ is the particle velocity, and $x_i(t)$ is the particle position with respect to iteration t. The parameters w, c_1, c_2, r_1, and r_2 are usually defined by the user. These parameter settings for this algorithm are usually constrained such that $w \in [0,1.2]$, $c_1c_2 \in [0,2]$, and $r_1r_2 \in [0,1]$. The term $wv_i(t)$ in Equation 4.1 (also referred to as the inertial term) maintains the particle's motion in the same direction as its original vector. The inertial coefficient w serves as a dampener or an accelerator during the movement of the particles. The term $c_1r_1\left[\hat{x}_i(t) - x_i(t)\right]$ better known as the cognitive component function serves as the memory. This component ensures that the particle tends to return to the position in the search space where the particle had a very high value of the fitness function. The term $c_2r_2[g(t) - x_i(t)]$ (also known as the social component) functions as mover of the particles to the position where the swarm has visited in the previous iterations. The best individual particle position, $\hat{x}_i(t)$, and the best swarm position, $g(t)$, for a minimization problem are computed as follows:

$$\hat{x}_i(t) = \min_{\tau \in [0,t]} x_i(t) \tag{7.7}$$

$$g(t) = \min_{i \in [1,n]} x_i(t) \tag{7.8}$$

where τ is the all the previous iterations from 0 to current iteration, t and n is the total number of particles. For a maximization problem the operators in Equations 7.7 and 7.8 are flipped to the maximization operator.

Next, the particles' position is then computed as is shown in the following:

$$x_i(t+1) = x_i(t) + v_i(t+1) \tag{7.9}$$

Until all candidate solutions are at their highest fitness positions and the termination criterion is satisfied, these iterations are then sustained. The execution scheme of the PSO technique is shown in Figure 7.1.

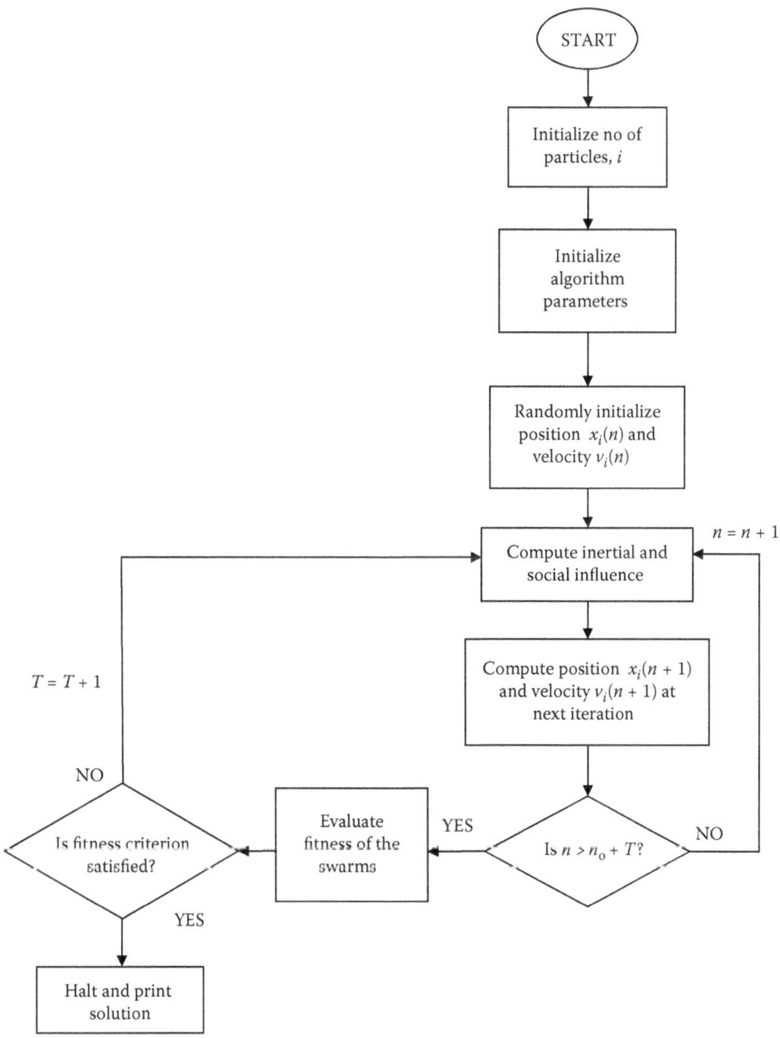

FIGURE 7.1 PSO flowchart.

7.3.3 Hopfield-Enhanced PSO

The HoPSO merges the ideas from the Ising spin models (Amit, Gutfreund, & Sompolinsky, 1986), the Hopfield neural networks (HNN) (Hopfield, 1982, 1984) and the PSO technique. The Ising model is a ferromagnetism model in statistical physics developed by Ernst Ising in 1925 (Dyson, 1969). This model is constructed based on the concept that atomic configurations can be represented in terms of magnetic dipole moments (atomic spins—the quantization of magnetism) which are in either in state +1 or −1. These spin interactions are localized to their closest neighbor and are usually in a lattice arrangement. The central idea of the Ising model is to detect phase transitions in real substances. Hence, repetitive magnetization would under certain circumstances and after a period of time cause the total energy of the magnetized material to converge into local minima.

The statistical physics models developed in the Ising model (Dyson, 1969) then inspired the development of a new type of neural net called the Hopfield recurrent artificial neural network (HNN) with improved convergence properties. The HNN was developed in 1982 by Hopfield (1982) and (1984). These neural nets observed to have applications in optimization problems (e.g., in Lee, Sode-Yome, & Park [1998] and Tank & Hopfield [1986]).

One of the key features of the HNN is that there is a decrease in the energy by a finite amount whenever there is a change in the network's state. This essential property confirms or accentuates the convergence of the output whenever the network state is changed. HNNs are usually constructed by a finite number of interlinked neurons. These neurons update their weights or their activation values (outputs from threshold neurons) independently relative to other neurons in the network. It is important to take note that the neurons in these sorts of networks are not directly connected to themselves and each neuron functions as an input as well as output. In HNNs, the activation values are usually binary (+1 or −1) and all the weights of neurons are symmetric ($w_{ij} = w_{ji}$).

The ideas of discreet magnetic spin as well as symmetric weight assumptions (from the Ising spin model) that were used in the HNN are applied to the PSO algorithm to improve its convergence capabilities. Similarly, a set of random weights w_{ij} was initialized and the symmetric property was imposed. Then, the modification of the particle position update equation in the PSO algorithm was done:

$$x_i(t+1) = x_i(t) + v_i(t) \tag{7.10}$$

where $x_i(t)$ is the particle position, $v_i(t)$ is the particle velocity, and t is the iteration counter. The weight previously defined is then introduced as a coefficient that dampens the position of the previous iteration such as the following:

$$s_j(t+1) = w_{ij}(t)x_i(t) + v_i(t) \tag{7.11}$$

The particle position at the next iteration is computed in the following piece-wise form similar to Equation 7.19:

$$x_j(t+1) = \begin{cases} +1 & \text{iff} \quad s_j(t+1) > U \\ -1 & \text{iff} \quad s_j(t+1) < U \\ x_j(t) & \text{iff} \quad \text{otherwise} \end{cases} \tag{7.12}$$

where U is the threshold which is user-defined. Then the energy function is defined as follows:

$$E = -\frac{1}{2}\sum_{j=1}\sum_{i=1} x_j(t+1)x_j(t)w_{ij} - \sum_{i=1} \theta x_j(t) \quad \text{such that} \quad j \neq i \tag{7.13}$$

where θ is the coefficient which is user-defined. This way the PSO algorithm is iterated until the total energy of the system reaches its local minima. To detect the instance when the HoPSO reaches the local minima, the difference between the energy levels at two consequent states is computed as follows:

$$dE = E_{n+1} - E_n \tag{7.14}$$

where n is the index that denotes the states. Thus, a new variant for the developed PSO algorithm is called the HoPSO approach. The algorithm and the flowchart for this HoPSO method are given in Algorithm 7.2 and Figure 7.2 respectively:

Algorithm 7.2: Hopfield Particle Swarm Optimization (HoPSO)

Step 1: Set no. of particles, i, and initialize parameter settings w, c_1, c_2, r_1, r_2, n_0
Step 2: Randomly initialize particles' position $x_i(t)$ and velocity $v_i(t)$
Step 3: Randomly initialize weights, $w_{ij}(t)$
Step 4: Enforce symmetry condition on weights
Step 5: Calculate cognitive and social components of the particles
Step 6: Compute position $x_i(t + 1)$ and velocity $v_i(t + 1)$ of the particles at next iteration
Step 7: Compute energy function
Step 8: Proceed with the evaluation of the fitness of each particle in the swarm
Step 9: If the energy difference between states are <0, proceed to Step 9 else go to Step 3
Step 10: If the fitness conditions are satisfied and $t < T_{max}$, stop program and print solutions, else go to Step 3.

7.4 THE CONCEPT OF NBI

NBI was first introduced and applied by Das and Dennis (1997). This method is a geometrically inspired scalarization technique for solving nonconvex MO problems. Unlike the weighted sum approach, the NBI method has the ability to find a

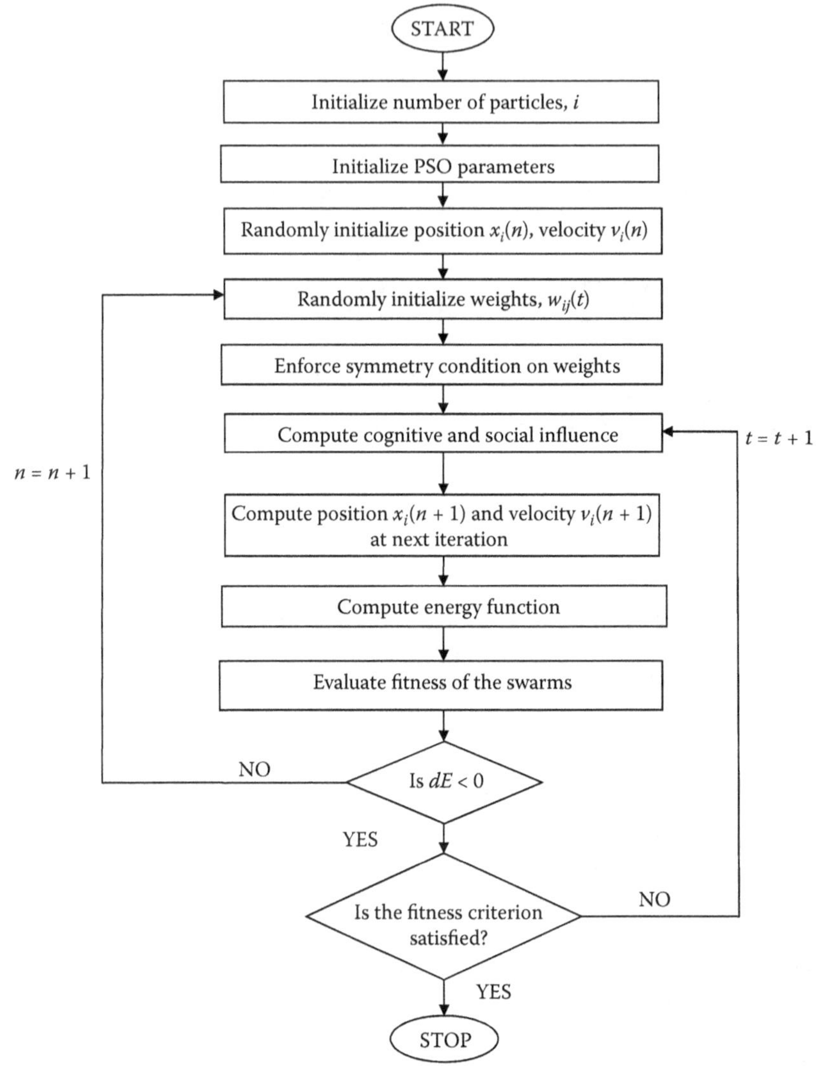

FIGURE 7.2 Flowchart for the HoPSO technique.

near-uniform spread of Pareto-optimal solutions in the frontier. This makes the NBI approach a more attractive option as compared to the weighted sum method when solving nonconvex, high-dimensional, and MO problems that consist of more than two objectives. Consider the generic MO problem as follows:

$$\text{Min}_{x \in X} F(x) \quad \text{s.t.}$$

$$X = \{x : g(x) = 0; h(x) \leq 0, x \in [1,4]\}$$

$$F^* = \left(f_i^* \; f_{i+1}^* \; .... \; f_{i+m}^*\right)^{\mathrm{T}} \tag{7.15}$$

where F^* is the utopia point for this MO problem. Let the individual minimum denoted as x_i^* be obtained for m number of objectives: $i \in [1,m]$. Thus, the convex hull of the individual minima is generated. The simplex from the convex hull is then presented as follows:

$$\Gamma = \left\{\phi \cdot Y : \phi = F(x_i^*); Y = \beta_i : i \in [1,m]\right\} \tag{7.16}$$

where ϕ forms a $m \times m$ matrix and $\sum_{i=1}^{m} \beta_i = 1$. The NBI β-subproblem is then formulated as follows:

$$\mathrm{Max}_{(X,t)} t \quad \text{s.t.}$$

$$\phi \cdot Y + tn = F(x); \quad x \in X \tag{7.17}$$

where t is some defined distance parameter, and n is the normal vector at the point toward the utopia point. Thus, the NBI scalarization finds the maximum distance, t, in the direction of the normal vector, n, between a point on the simplex and the origin (or the utopia point). Then the scalar values, Y, are varied systematically to develop a near-uniform spread of the Pareto frontier. The metaheuristic is then used to obtain the individual minima. Next, these algorithms are implemented to solve the subproblem for variable scalar values. By varying the scalar values, the results of the algorithms are used to construct the Pareto frontier.

Since the composed objective function in the problem statement involves both maximization and minimization, the NBI scheme cannot be directly applied. Thus, the objective function, f_3, was reformulated as an inverse ($1/f_3$). This way, the problem is now transformed to a purely maximization problem. The overall objective functions and constraints then are shown as follows:

$$\begin{array}{l}\text{Minimize} \rightarrow f_1, f_2, f_3 \\ \text{subject to operational constraints.}\end{array} \tag{7.18}$$

The procedures using which NBI concept is implemented are as follows:

1. The MO maximization problem is reformulated as a minimization problem by inverting the objective functions.
2. Obtain the local minima of individual objective functions.
3. The MO problem is then reformulated as a single-objective β-subproblem by using the value of the local optima.
4. The single-objective β-subproblem is then solved.

5. The inverse transformation is then performed to re-obtain the best maximal values of the objective functions.
6. The Pareto front for multiple scalarization values of β is obtained.
7. The optimal values of the solutions are then selected from the obtained Pareto front.

7.5 CONVERGENCE METRIC

The convergence metric (Deb & Jain, 2002) was developed to measure the degree of convergence of a solution set relative to a reference/basis set. Since the Pareto-optimal frontier was not known, a target vector P^*, which is the most dominant vector was employed as the reference set. For a set of solutions, the formulation to compute the convergence metric for a single run of the program is as follows:

$$d_i = \min_{j=1}^{|P^*|} \sqrt{\sum_{k=1}^{M} \left(\frac{f_k^i - f_k^j}{f_k^{\max} - f_k^{\min}} \right)} \tag{7.19}$$

where i and j denote the subsequent objective function values, k is the index which denotes the objective function, $f_k^{\max}$ is the maximum objective function value, $f_k^{\min}$ is the minimum objective function value, and M denotes the overall number of objectives. For this convergence metric, low metric values indicate high convergence characteristics among the solution vectors.

7.6 DETAILS ON THE COMPUTATIONAL EXPERIMENTS

The solution sets which construct the Pareto frontier are generated using metaheuristics within the NBI framework. The HVI and the convergence metric are utilized to evaluate and absolutely rank these solutions in terms of dominance. In these evaluations, the nadir point which is the most non-dominated point is utilized as a reference point for the computation of the hypervolume. The nadir point used is $(r_1, r_2, r_3) = (0, 0, 0)$. Along with the construction of the entire Pareto frontier, the individual solutions are classified into the best, median, and worst solutions. These solutions (which are generated by the metaheuristics) along with their respective ranks are shown in Table 7.1.

It can be observed in Table 7.1 that using the HVI, the best solution obtained by HoPSO algorithm dominates the best solution produced by the PSO algorithm and GA by 1.367% and 54.279%. The HVI computed for the entire frontier of each solution set produced by an algorithm gives the true levels of dominance when compared with another algorithm. The HVI for the entire frontier was computed for each of the algorithms. The level of dominance for the entire frontier for the solution sets produced by all the algorithms is shown in Figure 7.3.

The Pareto frontier produced by the HoPSO algorithm is more dominant than the ones produced by the GA and PSO algorithm by 75.106% and 0.497% respectively. The level of convergence of the Pareto frontier for all the algorithms employed is

TABLE 7.1
Individual Solutions Generated by the Metaheuristic Techniques

Metaheuristic	Description	Best	Median	Worst
HoPSO	Objective function (f_1, f_2, f_3)	(8.5675, 109.929, 24.8)	(8.54683, 109.903, 24.6649)	(8.56125, 108.326, 24.1797)
	Decision variable (x_1, x_2, x_3)	(55.6416, 72.8689, 33.5713)	(57.699, 72.5172, 36.9105)	(55.3454, 72.81, 22.7159)
	HVI	23,357.05	23,168.29	22,424.4
	Weights (w_1, w_2, w_3)	(0.5, 0.4, 0.1)	(0.1, 0.2, 0.7)	(0.5, 0.2, 0.3)
GA	Objective function (f_1, f_2, f_3)	(8.0301, 95.1255, 19.8195)	(7.7289, 91.1258, 18.606)	(7.6231, 88.7974, 17.7027)
	Decision variable (x_1, x_2, x_3)	(29.1553, 72.7404, 9.2592)	(23.8437, 72.6443, 9.9997)	(21.8074, 72.7049, 8.37861)
	HVI	15,139.5241	13,104.1788	11,983.0976
PSO	Objective function (f_1, f_2, f_3)	(8.53054, 108.941, 24.7943)	(8.5228, 108.861, 24.7803)	(8.52069, 108.842, 24.7694)
	Decision variable (x_1, x_2, x_3)	(49.6226, 72.8999, 30.6499)	(49.4068, 72.7734, 30.6636)	(49.4391, 72.6776, 30.5293)
	HVI	23,041.98	22,991.18	22,971.36
	Weights (w_1, w_2, w_3)	(0.5, 0.2, 0.3)	(0.3, 0.4, 0.3)	(0.2, 0.6, 0.2)

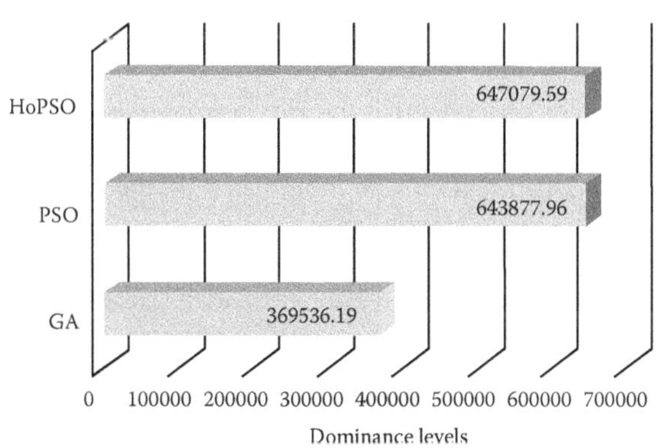

FIGURE 7.3 Level of dominance for the Pareto frontiers.

measured using the convergence metric (refer to Equation 7.26). The level of convergence of the Pareto frontiers generated by the GA, PSO, and the HoPSO techniques is shown in Figure 7.4.

The individual best solution of the HoPSO technique maximizes all the objectives f_1, f_2, and f_3 very effectively as compared to the solutions obtained by using

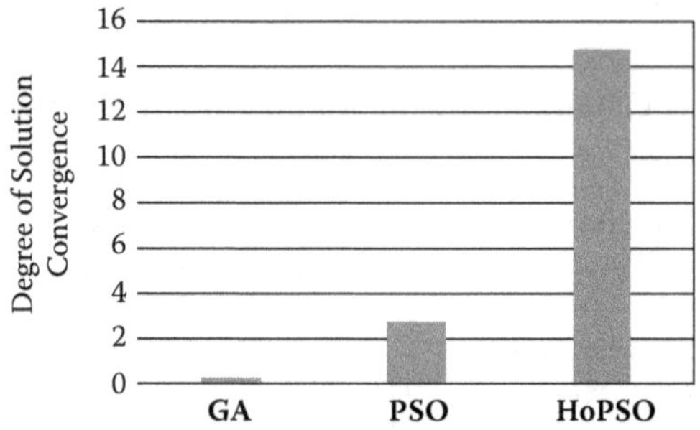

FIGURE 7.4 Level of dominance for the Pareto frontiers.

the GA and PSO algorithm (Table 7.1). Thus, it can be said that the HoPSO method outweighs the overall optimization capabilities of the PSO and GA techniques. The execution time for the GA, PSO, and the HoPSO algorithms to generate the entire frontier are 11.291, 13.113, and 13.268 seconds respectively. The HoPSO technique takes the longest computational time to produce the entire Pareto frontier followed by the PSO and GSA (gravitational search algorithm) methods. The HoPSO algorithm produces Pareto frontier with the highest level of dominance. However, due to high algorithmic complexity it sacrifices computational time the most as compared to the GA and PSO methods. The frontiers produced by the PSO and HoPSO algorithms are more uniform and diversely spaced in the objective space as compared to the GA. The frontier produced by the GA was confined to certain regions in the objective space. Thus, the solution spacing seems to influence heavily the ability of the algorithm to approximate the Pareto frontier dominantly. Localized solutions on the frontier such as the ones produced by the GA omit some of the solutions in the objective space. This in effect causes the GA to have a lower level of dominance as compared to the PSO and HoPSO algorithms.

In Figure 7.3, it can be observed that the Pareto frontier produced by the HoPSO method has a higher level of convergence as compared to the PSO technique. Figures 7.5 through 7.7 depict the frontier produced by the HoPSO, PSO, and GA techniques respectively in the objective space.

In this case, it could be inferred that the level of convergence as well as the level of dominance of the HoPSO algorithm are higher than those of the PSO technique. Therefore, in a post-analytical sense, the level of convergence may be one of the properties of the objective space which is interlinked to the level of dominance.

It can be observed in Table 7.1 that the HoPSO and PSO techniques share a similar feature during the optimization of the individual objectives. Both methods tend to emphasize on maximizing the second objective which is the geniposide yield as compared to the other objectives. In most computational optimization techniques, the factor of execution time usually conflicts with the notion of solution quality. This is because better solution quality is usually achieved when the technique spends more time searching the objective space thoroughly. However, in this case, the HoPSO technique seems to produce highly dominant solutions at the frontier and it simultaneously manages to do this with minimal execution time as compared to the GA approach. This may be attributed to the low algorithmic complexity of the HoPSO method as compared to the GA and most other metaheuristic approaches. The solutions constructing the Pareto frontier generated by the HoPSO technique is very diversely spread throughout the objective space as compared to the one produced by the other metaheuristics. The HoPSO technique effectively manages to spread its search to the region in the objective space with the highest fitness and not get deviated into other less fruitful search trajectories. This property gives the HoPSO technique more capability of finding more optimal solutions in the objective space.

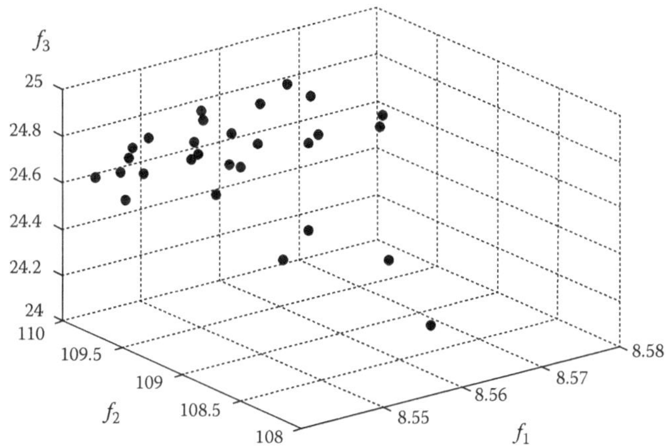

FIGURE 7.5 Pareto frontier of the objectives obtained by the HoPSO technique.

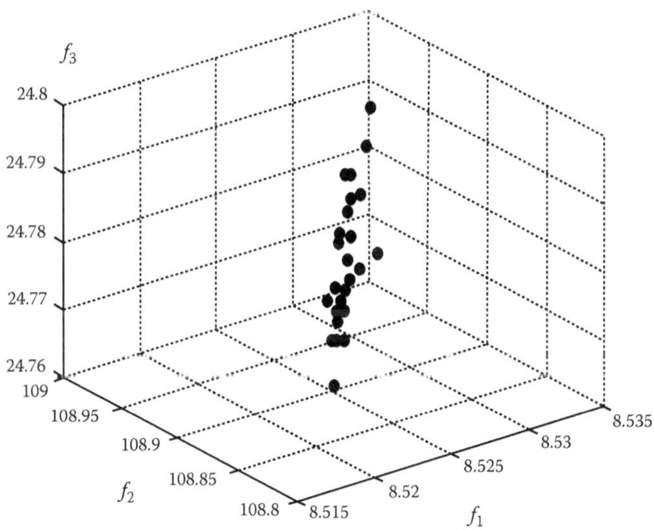

FIGURE 7.6 Pareto frontier of the objectives obtained by the PSO technique.

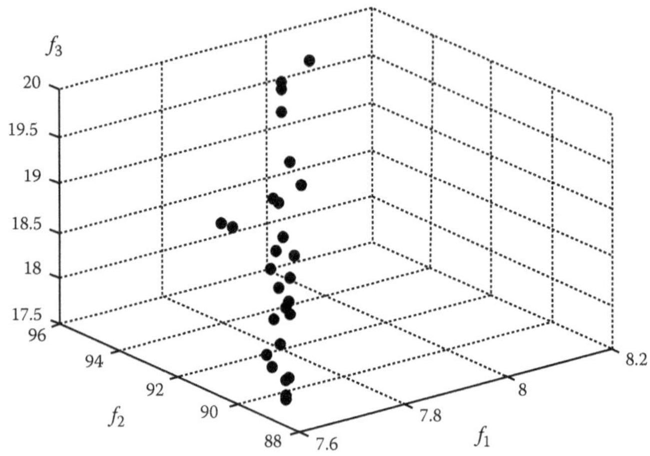

FIGURE 7.7 Pareto frontier of the objectives obtained by the GA technique.

The HVI is a reliable tool for ranking the solution spreads for a MO optimization problem. However, this metric is very dependent on the selection of the nadir point. If the correlation between the nadir point and the hypervolume value is irregular, then the HVI should be tested with multiple values of the nadir point to attain a trustworthy dominance ranking. Both evolutionary techniques employed in this work performed stable computations. Search stagnation, solution divergence, or halting problems did not occur during the numerical experimentations. The solutions constructing the Pareto frontier were all within the specified ranges and thus realistically feasible.

A new optimal set of solutions (see Pareto frontier in Figure 7.4) has been achieved using the HoPSO approach within the context of the NBI framework. This may be due to the geometrical aspect in the NBI framework which equips the metaheuristic with enhanced search capabilities as compared to the weighted sum approach which merely acts as a conventional scalarization system. Since HoPSO technique is an enhanced swarm-based algorithm (perturbative improvement), the diversification and the rigorousness of the search are high as compared with the GA and PSO methods.

7.7 MAIN CONCEPTS OF OPTIMIZATION PROCEDURES

There are several crucial insights gained from the optimization procedures conducted on the extraction process. The optimization resulted in obtaining a new optimal point and a dominant Pareto frontier using the HoPSO metaheuristic. The ranking of the metaheuristics based on their performance and the characteristics of their solution is a very effective procedure for choosing suitable metaheuristics for complex optimization problems. In addition, the metaheuristics have shown that there is still plenty of optimization potential in the second objective function. The HoPSO method outweighs the overall optimization capabilities of the GA and PSO techniques in terms of dominance levels. The level of convergence in the solution spread is highly interconnected to the degree of Pareto dominance. Due to this, the Hopfield enhancement in the HoPSO technique improves the generation of convergent solutions which in effect increases the dominance level of the solutions.

The swarm intelligence algorithms performed stable computations during the program execution and this technique generates feasible solutions that can be applied to Pareto frontier construction. Another advantage of swarm-based algorithms as compared to the other techniques is that it produces highly optimal results and is computationally very efficient. Gauging the solution quality using the HVI is a very useful approach when no information about the global optima or the existence of it is available. By obtaining the information regarding the level of dominance, approaches regarding further metaheuristic enhancement may be formulated and strategized.

REFERENCES

Aguirre, A. H., Zebulum, R. S., & Coello, C. (2004). Evolutionary multiobjective design targeting a field programmable transistor array. *Proceedings of the Conference on Evolvable Hardware, NASA/DoD* (pp. 199–205).

Amit, D. J., Gutfreund, H., & Sompolinsky, H. (1986). Spin-glass models of neural networks. *Physical Review A*, *32*(2), 1007–1018.

Das, I., & Dennis, J. E. (1997). A closer look at drawbacks of minimizing weighted sums of objectives for Pareto set generation in multicriteria optimization problems. *Structural Optimization*, *14*(1), 63–69.

Das, I., & Dennis, J. E. (1998). Normal-boundary intersection: A new method for generating the Pareto surface in nonlinear multicriteria optimization problems. *SIAM Journal of Optimization*, *8*(3), 631–657.

Deb, K., & Jain, S. (2002). Running performance metrics for evolutionary multiobjective optimization, KanGAL Report No. 2002004, Kanpur Genetic Algorithms Laboratory, Indian Institute of Technology, Kanpur, India.

Deb, K., Pratap, A., & Moitra, S. (2000). Mechanical component design for multiple objectives using elitist non-dominated sorting GA. In *Proceedings of the Parallel Problem Solving from Nature VI Conference*, Paris (pp. 859–868).

Deb, K., Pratap, A., Agarwal, S., & Meyarivan, T. (2002). A fast and elitist multiobjective genetic algorithm: NSGA-II. *IEEE Transactions on Evolutionary Computation*, *6*(2), 182–197.

Dyson, F. J. (1969). Existence of a phase-transition in a one-dimensional Ising ferromagnet. *Communications in Mathematical Physics*, *12*, 91–107.

Emmerich, M., Beume, N., & Naujoks, B. (2005). An EMO algorithm using the hypervolume measure as selection criterion. In *International Conference on Evolutionary Multi-Criterion Optimization (EMO 2005)*, Guanajuato, Mexico, 9–11 March (pp. 62–76). Berlin, Heidelberg: Springer.

Fishburn, P. C. (1967). Additive utilities with incomplete product set: Applications to priorities and assignments. *Operations Research*, *15*, 537–542.

Gunda, J., & Acharjee, P. (2011). Multiobjective economic dispatch using Pareto frontier differential evolution. *International Journal of Engineering Science and Technology*, *3*(10), 7389–7396.

Heo, J. S., & Lee, K. Y. (2006). Multiobjective control of power plants using particle swarm optimization techniques. *IEEE Transactions on Energy Conversion*, *21*(2), 552–561.

Hopfield, J. J. (1982). Neural networks and physical systems with emergent collective computational abilities. *Proceedings of National Academy of Sciences*, *29*, 2554–2558.

Hopfield, J. J. (1984). Neurons with graded response have collective computational properties like those of two-state neurons. *Proceedings of National Academy of Sciences, 81*, 3088–3092.

Kennedy, J., & Eberhart, R. (1995). Particle swarm optimization. *IEEE Proceedings of the International Conference on Neural Networks* (pp. 1942–1948). Perth, Australia.

King, R. T. A., Rughooputh, H. C., & Deb, K. (2005). Evolutionary multi-objective environmental/economic dispatch: Stochastic versus deterministic approaches. In *International Conference on Evolutionary Multi-Criterion Optimization* (pp. 677–691) March. Berlin, Heidelberg: Springer.

Kusiak, A., Xu, G., & Tang, F. (2010). Optimization of an HVAC system with a strength multi-objective particle-swarm algorithm. *Energy, 36*(5), 184–192.

Lee, K. Y., Sode-Yome, A., & Park, J. H. (1998). Adaptive Hopfield neural networks for economic load dispatch. *IEEE Transactions on Power Systems, 13*(2), 519–526.

Li, H. B., Wong, C. C., Chen, K. W., & Chen, F. (2008). Antioxidant properties in vitro and total phenolic contents in methanol extracts from medicinal plants. *LWT Food Science and Technology, 41*, 385–390.

Luyben, M. L., & Floudas, C. A. (1994). Analyzing the interaction of design and control. 1. A multiobjective framework and application to binary distillation synthesis. *Computers and Chemical Engineering, 18*, 10, 933–969.

Nain, P. K. S., Giri, J. M., Sharma, S., & Deb, K. (2010). *Multi-objective performance optimization of thermo-electric coolers using dimensional structural parameters* (pp. 607–614). SEMCCO 2010, LNCS 6466. Berlin, Heidelberg: Springer-Verlag.

Neshat, M., Sepidnam, G., Sargolzaei, M., & Toosi, A. N. (2014). Artificial fish swarm algorithm: A survey of the state-of-the-art, hybridization, combinatorial and indicative applications. *Artificial Intelligence Review, 42*(4), 965–997.

Oshima, T., Sagara, K., Yoshida, T., Tong, Y. Y., Zhang, G., & Chen, Y. H. (1988). Determination of geniposide, gardeniside, geniposidic acid and genipin-1-β-gentiobioside in *Gardenia jasminoides* by high-performance liquid chromatography. *Journal of Chromatography A, 455*, 410–414.

Palli, N., Azram, S., McCluskey, P., & Sundararajan, R. (1999). An interactive multistage ε-inequality constraint method for multiple objectives decision making. *ASME Journal of Mechanical Design, 120*(4), 678–686.

Rashedi, E., Nezamabadi-pour, H., & Saryazdi, S. (2009). GSA: A gravitational search algorithm. *Information Sciences, 179*, 2232–2248.

Reddy, M. J., & Kumar, D. N. (2007). An efficient multi-objective optimization algorithm based on swarm intelligence for engineering design. *Engineering Optimization, 39*(1), 49–68.

Sandgren, E. (1994). Multicriteria design optimization by goal programming. In H. Adeli (Ed.), *Advances in design optimization* (Chapter 23, pp. 225–265). London: Chapman & Hall.

Shashi, Deep, K., & Katiyar, V. K. (2010). Multiobjective extraction optimization of bioactive compounds from Gardenia using real coded genetic algorithm. *6th World Congress of Biomechanics (WCB 2010), IFMBE Proceedings* (Vol. 31, pp. 1463–1466), Berlin: Springer.

Song, Z., & Kusiak, A. (2010). Multiobjective optimization of temporal processes. *IEEE Transactions on Systems, Man, and Cybernetics, Part B (Cybernetics), 40*(3), 845–856.

Statnikov, R. B., & Matusov, J. B. (1995). *Multicriteria optimization and engineering*. New York: Chapman and Hall.

Tank, D. W., & Hopfield, J. J. (1986). Simple neural optimization network: An A\D converter, signal decision circuit and a linear programming circuit. *IEEE Transactions on Circuits and Systems*, CAS-33, 533–541.

Triantaphyllou, E. (2000). *Multi-criteria decision making: A comparative study* (p. 320). Dordrecht, The Netherlands: Kluwer Academic Publishers (now Springer).

Van Calsteren, M. R., Bissonnette, M. C., Cormier, F., Dufresne, C., Ichi, T., LeBlanc, J. C. Y., et al. (1997). Spectroscopic characterization of crocetin derivatives from *Crocus sativus* and *Gardenia jasminoides*. *Journal of Agricultural and Food Chemistry*, *45*, 55–1061.

Van Sickel, J. H., Venkatesh, P., & Lee, K. Y. (2008). Analysis of the pareto front of a multi-objective optimization problem for a fossil fuel power plant. In *Power and Energy Society General Meeting-Conversion and Delivery of Electrical Energy in the 21st Century*, July, 2008, (pp. 1–8). Piscataway, NJ: IEEE.

Yang, B. S., Yeun, Y. S., & Ruy, W. S. (2002). Managing approximation models in multiobjective optimization. *Structural Multidisciplinary Optimization*, *24*, 141–156.

Zitzler, E., & Thiele, L. (1998). Multiobjective optimization using evolutionary algorithms—A comparative case study. In *Conference on Parallel Problem Solving from Nature (PPSN V)*, Amsterdam, the Netherlands (pp. 292–301). Berlin, Heidelberg: Springer.

8 Multiobjective Optimization of Cement-Bonded Sand Mold System

In the manufacturing sector and other heavy industries, problems encountered by engineers and decision makers often present themselves in a multiobjective (MO) setting (Ganesan, Vasant, & Elamvazuthi, 2011; Ganesan, Vasant, Elamvazuthi, & Shaari, 2012). Due to stringent environmental regulations and tight energy requirements, there exists a high demand for obtaining efficient strategies to optimize operations in these industries. Thus, to effectively optimize these systems, the decision maker is required to have a comprehensive understanding of computational intelligence and material engineering.

8.1 OPTIMIZING SAND MOLD CASTING PROCESSES

Sand casting is one of the earliest forms of casting where cavities are introduced into the sand and then liquid metal is poured into the cavity. Since oiled sand molds pack well and have the capability to hold their shape, they are often employed for metal casting. The liquid metal then takes the shape of the cavity and forms the metal product once cooled. Even today, sand casting remains the cheapest form of casting due to its simplicity and availability of the materials. In addition, sand has very suitable physical properties such as permeability, strength, cohesiveness, refractoriness, adhesiveness, flowability, and collapsibility.

Due to this fact, many research efforts have been directed to optimizing the sand casting process. For instance, Kumar and Singh (2011) carried out a design-of-experiments approach using the Taguchi technique (Guharaja, Noorul Haq, & Karuppannan, 2006). In Kumar and Singh (2011), three manipulated parameters, the grain fineness number, clay content, and moisture content, were selected and optimized using the Taguchi technique (Guharaja et al., 2006) to obtain the maximum tensile strength. In their work, the clay content was found to have the most significant effect on the tensile strength. Thus, they concluded that higher clay content and grain fineness number along with a medium-level moisture content lead to optimal tensile strength. In Keswani and Upadhye (2012), similarly, the optimization of the casting process to minimize casting defects was carried out. The optimization strategy was directed to identify optimal factors affecting the casting process by running a small number of experiments. The Taguchi technique was employed to maximize the signal-to-noise ratios and to minimize the noise factors. Using this strategy the authors managed to reduce casting defects to a significant level.

In the past, the Taguchi technique has been widely applied for optimizing mold casting processes. Another interesting application involving this technique was carried out in the work of Kim, Shin, Lee, and Moon (2007). In that work, the correlations between the process parameters and the mechanical properties of the product were investigated using the Taguchi technique on 14.5% silicon-containing corrosion-resistant cast iron. The mechanical properties were analyzed using fractography studies, microscopic test, hydrogen content studies, and acid resistance test.

Besides the Taguchi approach, metaheuristics such as particle swarm optimization (PSO) have been implemented to certain types of casting problems. In Ji and Jiang (2012), an optimization tool was developed to improve the cooling conditions for the spray zones of a real continuous caster. This optimization tool consists of a two-dimensional heat transfer model and the adaptive evolutionary PSO technique. By using the adaptive evolutionary PSO, the authors managed to obtain optimal control of the secondary cooling water distribution according to certain metallurgical criteria and operational constraints. In Xudong, Zhaofeng, and Man (2011) PSO has also been used to approximate mold oscillation status during the continuous casting process. In that work, the PSO was implemented effectively and the effects of the variations in the algorithmic parameters on the oscillation approximation were studied. The PSO was successfully applied in the work of Xudong et al. (2011) to solve the inverse estimation for the mold oscillation status (e.g., stiffness, empty driving, and dampness).

Recently, genetic algorithms (GAs) have also been used for optimizing the continuous casting process. In Bhattacharya, Debjani, Roychowdhury, and Das (2007), the optimum selection of mold oscillation parameters for continuous casting in steel manufacturing was carried out using GA. The main aim of that work was to improve the lubrication inside the mold so as to reduce the intensity of oscillation marks during high casting speeds. The optimization in Bhattacharya et al. (2007) was done by implementing the GA to improve the frequency, amplitude, and waveform of the oscillation during the continuous casting in a way that it maximizes the lubrication, minimizes the depth of oscillation marks, and minimizes the cycle peak friction. The constraints used during the optimization were the machine operational bounds. As compared to the original parameters selected using the conventional/heuristic approaches set by the original equipment manufacturer, the selection done using GA by Bhattacharya et al. (2007) has been shown to be more effective in optimizing the individual oscillation effects. Another interesting application of GAs in the casting system was carried out by Lagdive and Inamdar (2013). In that work, the optimization of the design of the riser or feeder was done. Risers are incorporated into the setup to feed liquid metal into casting which is solidified to compensate for solidification shrinkage. Therefore, in Lagdive and Inamdar (2013), by utilizing the GA technique, the optimal riser size was identified for each of the casting components. It was also found that the GA technique could be applied to standard feeder shape since the geometric formulation can be used to calculate the volume and the area. An illustration of the optimization scenario in sand mold casting is shown in Figure 8.1.

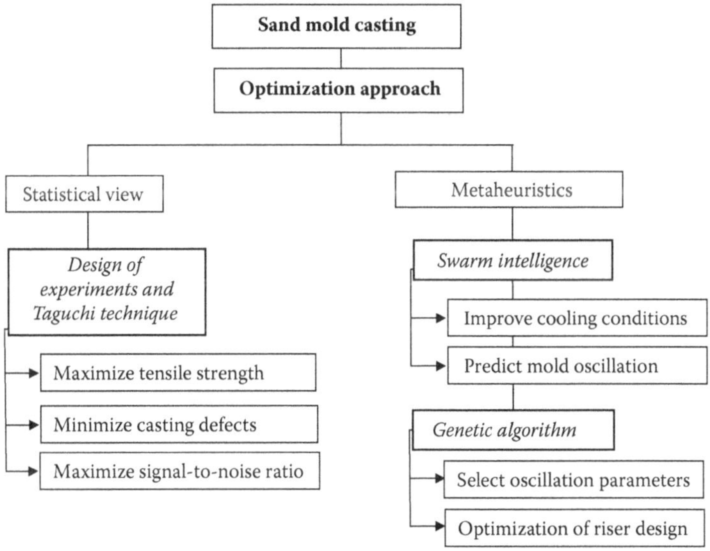

FIGURE 8.1 Optimization scenario of sand mold casting.

8.2 MO OPTIMIZATION OF CEMENT-BONDED SAND MOLD SYSTEM

The MO optimization of the cement-bonded sand mold system is presented in this chapter. The main idea behind these sorts of molds is that a binder element (in this case cement) is used to enhance the mold strength, mold hardness, and the casting accuracy. The primary constituents of the cement bonded sand mold system are cement, silica sand, and water (Surekha, Vudavilli, & Parappagoudar, 2012). However, a lot of time is taken for the production of cement-bonded sand molds that respect the strength and hardness requirements. To rectify this issue, a constituent called an "accelerator" (which is usually calcium formate) is introduced.

The cement-bonded sand mold system problem presented in Surekha et al. (2012) was tackled using two metaheuristic techniques. Differential evolution (DE) (Storn & Price, 1995) is a population-based evolutionary algorithm that has been derived from GAs (Holland, 1992). DE was developed in the 1990s and has been used extensively to solve problems which are non-differentiable, noncontinuous, nonlinear, noisy, flat, multidimensional, have many local minima and constraints, or have high degree of stochasticity. DE has been widely applied in many areas ranging from engineering and economics to operational research (Babu & Angir, 2003; Das, Abraham, & Konar, 2008; Ali, Pant, & Abraham, 2009).

In addition, a new variant of the DE technique which utilizes concepts from Hopfield networks (Hopfield, 1982, 1984; Vasant, Ganesan, & Elamvazuthi, 2012) was developed. This variant was developed with an aim to improve the convergence properties of the DE algorithm and by implication improving its overall efficiency in attaining optimal solutions. This improved technique called the Hopfield differential evolution (Hopf-DE) algorithm was then tested and compared with the standard DE algorithm in this work. Both optimization techniques employed were used in conjunction with the weighted sum scalarization approach (Fishburn, 1967; Triantaphyllou, 2000) to generate a series of solutions that dominantly approximate the Pareto frontier. Analysis was then conducted to identify the individual best, median, and worst solutions as well as the frontier approximations obtained in this work. The details of the formulation employed to optimize the cement-bonded mold system are presented. The optimization strategy is discussed and an in-depth explanation of the metaheuristic mechanism is provided. The process of measuring the performance of the metaheuristic algorithm during the optimization is analyzed.

8.3 MODELING CEMENT-BONDED SAND MOLD SYSTEMS

In the cement-bonded sand mold, cement is used as a binder to produce highly accurate casting. Besides, using cement as a binder element also causes the mold core to have high compressive strength. The three critical constituents in the cement-bonded sand mold are sand, water, and cement. The major constituents of cement required are dicalcium silicate ($2CaO{\cdot}SiO_2$), tricalcium aluminate ($3CaO{\cdot}Al_2O_3$), tricalcium silicate ($3Ca{\cdot}SiO_2$), and tetracalcium aluminoferrite ($4CaO{\cdot}Al_2O_3{\cdot}Fe_2O_3$). To produce a mold with high compressive strength, the total casting time would have to be very long. Therefore, accelerators are often employed to reduce the casting time without sacrificing the compressive strength of the mold by increasing the rate of hydration. An accelerator is a chemical constituent that complies with the American Society for Testing Materials (ASTM) Type C standard. Examples of commonly used accelerators in concrete mixtures are calcium chloride, calcium nitrate, sodium chloride, calcium nitrite, calcium aluminate, sodium formate, and calcium formate (Neville, 2000). The cement-bonded sand mold model considered in this chapter was developed by Surekha et al. (2012). In that work, the accelerator utilized was calcium formate ($Ca[OOOCH]_2$). Hydration also increases the resistance of the mold to chemical reactions (Sleicher & Providence, 1943).

In the past, very few works have been conducted on the modeling of the physical properties of the cement-bonded molds relative to their input responses (e.g., amount of cement, water content, etc.). The work of Uchikawa and Uchida (1976) has provided useful information regarding the responses of compressive strength with respect to the amount of water, amount of jet cement, duration of testing time, and the amount of accelerator. Their rigorous studies have resulted in the identification of the critical parameters that critically influence the compressive strength of the sand mold. The utilization of the statistical viewpoint has been demonstrated in Montgomery (2001) to be capable of generating the response models of the cement-bonded sand mold system. The statistical view highlighted is the design-of-experiments approach encompassing efficient data collection and regression modeling of the input–output relations. Another excellent approach for ascertaining the nonlinear response relations of the cement-bonded system was used in Parappagoudar, Pratihar, and Datta (2007) by utilizing the response surface methodology (RSM) in tandem with the design-of-experiments approach. The RSM approach uses polynomial and localized estimations from empirical data to obtain the linear response relations.

The combination of the two approaches allows the modeling to be done accurately and efficiently with minimal amount of experimental rigor while capturing all the critical features of the system involved. However, due to the low accuracy of the linear model when it comes to nonlinear responses, the authors in Parappagoudar et al. (2007) employ the central composite design (CCD) framework (Montgomery, 2001) to construct the nonlinear response models. In Parappagoudar et al. (2007), an accelerator was employed to speed up the cement-bonded sand mold development. Experiments were then conducted where the data obtained were employed to construct linear models for responses of the compressive strength and mold hardness. These linear models were then tested using randomized statistical tests. Consequently, the nonrotatable CCD (Anderson & Whitcomb, 2005) was used to develop the nonlinear models for the mentioned responses. Similarly, the robustness of these nonlinear models was gauged using randomized statistical tests. The authors of Parappagoudar et al. (2007) then performed comparative studies between their linear and nonlinear models with the aid of some randomly developed test cases. They then identified the best model which had the minimal percentage of deviation for the prediction of each responses. Finally, they successfully completed the work by developing the relationship between the responses of the cement-bonded mold system.

Therefore, the process parameters of the mold heavily influence the quality of the final product. In Surekha et al. (2012), these parameters were identified as the decision variables while the hardness (H) and the compression strength (CS) of the mold were used as the objective functions. The process parameters were the ratio in terms of percentage of Portland cement (A), percentage of accelerator (B), percentage of water content (C), and the testing time in hours (D). The objective functions and the range of the decision variables are shown as follows:

$$\begin{aligned} H = {} & 13.5199 + 7.38194A + 11.6111B + 6.76042C + 0.113812D \\ & - 0.388889A^2 - 1.68056B^2 - 9.22222C^2 + 0.0493827D^2 \\ & - 0.289062AB + 1.57813AC + 0.0130208AD + 1.59375BC \\ & + 0.0989583BD + 0.30208CD \end{aligned} \tag{8.1}$$

$$\begin{aligned} \text{CS} = {} & 83.8913 - 3.59381A - 2.45386B + 135.424C - 18.6127D \\ & + 0.213865A^2 - 3.28139B^2 - 22.7782C^2 + 1.39739D^2 \\ & + 2.75252AB - 4.12608AC + 2.61247AD - 4.07222BC \\ & + 2.12947BD - 7.73506CD \end{aligned} \tag{8.2}$$

$$\begin{aligned} & 8 \le A \le 12 \\ & 2 \le B \le 4 \\ & 3 \le C \le 8 \\ & 2 \le D \le 8 \end{aligned} \tag{8.3}$$

The MO optimization problem statement for the green mold system problem is shown as follows:

$$\begin{aligned} &\text{Max}\left(f_1, f_2\right) \quad \text{subject to} \\ &8 \le A \le 12 \\ &2 \le B \le 4 \\ &3 \le C \le 8 \\ &2 \le D \le 8 \end{aligned} \tag{8.4}$$

8.4 BACKGROUND OF THE WEIGHTED SUM APPROACH

In MO optimization, the weighted sum approach, and other scalarization techniques are considered to be classical approaches. "Classical" is used in the sense that they precede most of the more modern metaheuristic approaches such as evolutionary and swarm-based techniques. However, currently, classical approaches are often employed in tandem with metaheuristics since they provide a strong foundation to the overall computational strategy (Elamvazuthi, Ganesan, & Vasant, 2011; Ganesan, Elamvazuthi, Shaari, & Vasant, 2013).

One useful way to classify MO optimization is to divide it into two groups which are the preference-based and ideal procedures. The preference-based procedures usually involve the aggregation of objective functions to form a single composite function. This function is the weighted total of the individual objective functions. This composite function is then maximized or minimized depending on the problem definition. This approach is very useful if the relative preference of the objectives are known prior to the optimization procedures. The relative preferences are then represented as the weights which would be utilized to aggregate the individual objectives.

The ideal procedure revolves around the idea of finding multiple (non-weighted) solutions rigorously for a large range of values in the decision variables. For the selection of the desired solution, this approach requires the decision maker to have access to higher-level knowledge to aid with design of the criterion (approximate function or vector) employed for the selection. The ideal procedure is non-preference method that does not use any preference information about any of the objective functions. Non-preference methods make use of some form of heuristic or utility function to identify a single optimal solution. This approach is usually computationally intensive and requires a thorough search of the objective space prior to obtaining the desired solution option. Preference-based procedures can be further grouped into three classes which are the posteriori, a priori, and interactive methods. Posteriori methods are carried out by employing preference information about the objective functions to produce a Pareto-optimal solution set.

A priori methods use preference information about the objective function to identify a single Pareto-optimal solution while interactive procedures involve the employment of preference information throughout the execution of the optimization procedures. The classification of these procedures is graphically presented in Figure 8.2.

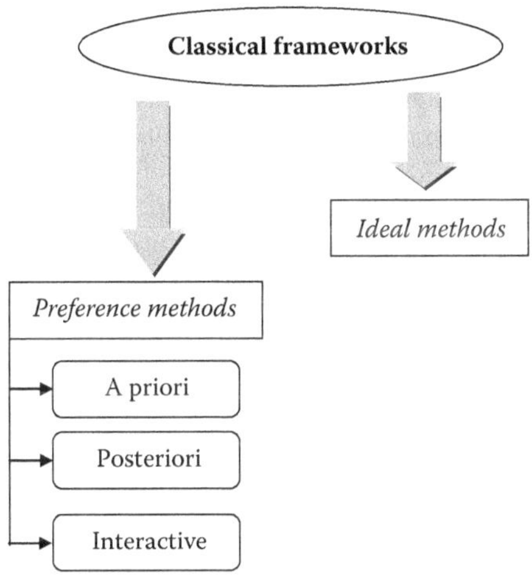

FIGURE 8.2 Classification of classical MO approaches.

The weighted sum approach falls under the subclass of preference methods since relative importance/preference values are assigned to the individual objective functions prior to aggregation. The weighted sum approach is the most widely used approach in the past and is still highly relevant in today's optimization scene. Its simplicity and its strong mathematical foundation make it highly suitable to be applied per se or to be implemented hand in hand with other more modern techniques such as metaheuristics.

8.5 DIFFERENTIAL EVOLUTION

DE is a class of evolutionary metaheuristic algorithms first introduced in 1995 by Storn and Price. The core idea of this technique is the incorporation of perturbative methods into evolutionary algorithms. Before diving into the technical details of the DE algorithm, a brief overview on some of the applications of DE in material engineering is provided in the subsequent paragraphs.

In material engineering and industrial manufacturing, DE has become increasingly popular due to its robustness, effectiveness, and ease of use. For instance, in Bhattacharya and Sambasivam (2009), DE was employed for the optimization of the oscillation parameters in the continuous casting process of steel manufacturing. In this casting process, liquid steel is poured into a buffer vessel and is consequently flowed into a water-cooled copper mold via a nozzle. The water is circulated to cool the liquid steel. This continuous casting process consists of two other processes namely mold oscillation and lubrication addition. These processes are targeted to reduce the sticking of the formative steel shell to the mold and nonuniformity of the shell development due to irregular heat transfer. In Bhattacharya and Sambasivam (2009), the authors develop a mathematical formulation to express the correlation between the oscillation of the free parameters and their respective responses. This formulation is then used to develop an optimization model to maximize the lubrication and minimize the depth of the oscillation by manipulating the oscillation frequency, oscillation amplitude/stroke, and sinusoidal variation. Therefore, the authors in Bhattacharya and Sambasivam (2009) are faced with a bi-objective optimization problem. The primary goal of that work was to employ DE and GA to solve the formulated optimization problem and to see both algorithms return the same solutions for various casting speeds. Second, the authors compared their algorithmic efficiencies with respect to convergence characteristics and computational time. It was found that the DE approach yielded the same quality of results as compared to the GA approach. However, the DE approach has a higher convergence rate as compared to the GA approach.

The work of Jawaha and Ramamoorthy (2012) involves the dynamic optimization of the injection molding using three evolutionary approaches (DE, GA, and PSO). Injection molding is a highly nonlinear, nonconvex, and multivariate procedure employed for processing plastics from thermoplastic materials. Using this procedure, thermoplastic material can then be molded into complex geometrical shapes. An injection molding process consists of four different mechanisms which are plasticizing, injection, clamping, and ejection. The plasticizing mechanism melts the resin material to improve flowability into the mold. Prior to injection, the resin is mixed and melted in a temperature-controlled barrel. Next, the clamping unit adjusts the opening and closing of the mold to maintain a steady force on the mold during the injection. The ejector pin ejects the final mold product from the die. The injection molding cycle is heavily influenced by the processing conditions such as injection pressure, back pressure, barrel temperature, nozzle temperature, holding pressure, clamping force, cooling time, filling time, and screw rotating speed. The authors in Jawaha and Ramamoorthy (2012) optimized the mentioned process parameters online so as to minimize the product defects by employing PSO, DE, and GA. The authors found that all metaheuristics performed well with the exception of PSO which undergoes fast convergence and generates unsatisfactory solutions as compared to GA and DE.

Another interesting application of DE in material engineering can be seen in the work done in Holis, Skanderova, Placek, Dvorsky, and Zelinka (2013). In that work, the DE and the self-organizing migration algorithm (SOMA) was utilized to solve the steel alloying problem. The steel alloying problem involves manipulation of the weight of input steel, composition of input steel, list of alloying materials (and their prices per unit), and the target steel grade. These optimal settings of these parameters are identified to maximize the quality of the manufactured steel (i.e., final steel grade) at a reduced per-unit price with minimal energy consumption. The DE and the SOMA techniques were found to be effective in solving the problem as compared to the Simplex method. The authors in Holis et al. (2013) found that although the Simplex approach may produce the cheaper solution, the implementation of its metaheuristics struck a fine balance between steel grade, price per unit, and energy consumption.

The DE technique starts by the initialization of a population of at least four individuals denoted as P. These individuals are real-coded vectors with some size N. The initial population of individual vectors (the first generation denoted as gen = 1) is randomly generated in appropriate search ranges. One principal parent denoted x_i^{p} and three auxiliary parents denoted x_i^{a} are randomly selected from the population, *P*. In DE, every individual in the population, *P*, would become a principal parent, x_i^{p}, at one generation or the other and thus have a chance in mating with thc auxiliary parents, x_i^{a}. The three auxiliary parents then engage in "differential mutation" to generate a mutated vector, V_i.

$$V_i = x_1^{\mathrm{a}} + F\left(x_2^{\mathrm{a}} - x_3^{\mathrm{a}}\right) \tag{8.5}$$

where *F* is the real-valued mutation amplification factor which is usually between 0 and 1.

Next V_i is then recombined (or discreetly crossed over) with x_i^{p} to generate child trial vector, x_i^{child}. The crossover operator is defined as follows:

$$CR : V_i \otimes x_i^{\mathrm{p}} \rightarrow x_i^{\mathrm{child}} \tag{8.6}$$

The probability of the crossover, *CR*, is the probability of amount of codes from the mutated vector, V_i, with respect to the principal parent, x_i^{p} being inherited by the child trail vector. The *CR* is an input parameter set by the user. In DE, the survival selection mechanism into the next generation is called "knock-out competition." This is defined as the direct competition between the principal parent, x_i^{p}, and the child trial vector, x_i^{child}, to select the survivor of the next generation as follows:

$$x_i(\text{gen}+1) = \begin{cases} x_i^{\mathrm{child}}(\text{gen}) \leftrightarrow f\left(x_i^{\mathrm{child}}\right) \text{ better than } f\left(x_i^{\mathrm{p}}\right) \\ x_i^{\mathrm{p}}(\text{gen}) \leftrightarrow \text{otherwise} \end{cases} \tag{8.7}$$

Therefore, the knock-out competition mechanism also serves as the fitness evaluation scheme for the DE algorithm. The DE method is described in Algorithm 8.1 and Figure 8.3.

Algorithm 8.1: Differential Evolution (DE)

Step 1: Initialize parameters *N*, *P*, *CR*, and *F*
Step 2: Randomly initialize the population vectors, x_i^{G}
Step 3: Randomly select one principal parent, x_i^{p}
Step 4: Randomly select three auxiliary parents, x_i^{a}
Step 5: Perform differential mutation & generate mutated vector, V_i
Step 6: Recombine V_i with x_i^{p} to generate child trial vector, x_i^{child}
Step 7: Perform "knock-out" competition for next-generation survival selection
Step 8: If the fitness criterion is satisfied and $t = T_{\max}$, halt and print solutions, else proceed to step 3

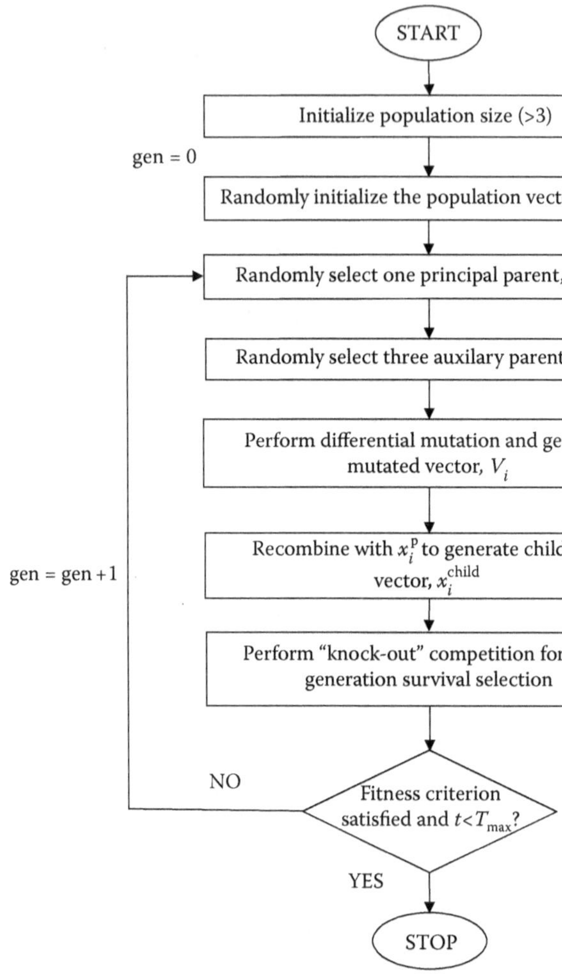

FIGURE 8.3 Flowchart of the DE algorithm.

8.6 HOPFIELD DE

The central ideas used during the design of the Hopf-DE spring from notions first introduced in Ising spin models (Amit, Gutfreund, & Sompolinsky, 1986) and the Hopfield neural networks (HNN) (Hopfield, 1982, 1984). The statistical physics models developed in the Ising model (Dyson, 1969) then inspired the development of a new type of neural net called the Hopfield recurrent artificial neural network with improved convergence properties. The HNN developed by Hopfield (1982, 1984) was observed to have applications in optimization problems (Tank & Hopfield, 1986; Lee, Sode-Yome, & Park, 1998). One of the key features of the HNN is that there is a decrease in the energy by a finite amount whenever there is a change in the network's state. This essential property confirms or accentuates the convergence of the output whenever the network state is changed.

HNNs are usually constructed by a finite number of interlinked neurons. These neurons update their weights or their activation values (outputs from threshold neurons) independently relative to other neurons in the network. It is important to take note that the neurons in these sorts of networks are not directly connected to themselves and each neuron functions as an input as well as output. In HNNs, the activation values are usually binary (+1 or −1) and all the weights of neurons are symmetric ($w_{ij} = w_{ji}$). Since HNNs are back-propagating auto-associator networks, the input of the neurons is updated by the activation values of the output. Therefore, the input of a neuron i at a cycle t+1 is as follows:

$$s_i(t+1) = \sum_{j=1} \sum_{i=1} y_j(t) w_{ji} + \theta \quad \text{such that} \quad \forall j \neq i \tag{8.8}$$

where $y_j(t)$ is the activation value from the output, w_{ji} is the neural weight, and θ is some arbitrary constant set by the user. A piece-wise threshold function is applied to the neural inputs to obtain the activation values of the outputs, $y_j(t+1)$, at the next cycle $(t+1)$:

$$y_j(t+1) = \begin{cases} +1 & \text{iff} \quad s_i(t+1) > U \\ -1 & \text{iff} \quad s_i(t+1) < U \\ y_j(t) & \text{iff} \quad s_i(t+1) = U \end{cases} \tag{8.9}$$

Therefore

$$y_j(t+1) = \text{sgn}(s_i(t+1)) \tag{8.10}$$

where U is some user-defined threshold value. A neuron i in the HNN is termed "stable" if the following condition is satisfied:

$$y_j(t) = \text{sgn}(s_i(t-1)) \tag{8.11}$$

In HNN systems, an additional symmetry restriction on the weights ($w_{ij} = w_{ji}$) is imposed. This is done so that the neurons are bidirectional. Hence, the network has the capability to settle to a stable state and this enforces the convergent behavior of the algorithm. With this additional restriction, the energy function of the system can be defined as follows:

$$E = -\frac{1}{2} \sum_{j=1} \sum_{i=1} y_j y_i w_{ji} \quad \sum_{i=1} \theta y_i \quad \text{such that} \quad j \neq i \tag{8.12}$$

Therefore, after a finite number of steps the network converges into a stable state and the energy reaches its local minima (analogous to the atomic energy of the simulated annealing (SA) algorithm (Metropolis, Rosenbluth, Rosenbluth, Teller, & Teller, 1953; Kirkpatrick, Gelatt, & Vecchi, 1983; Ingber, 1993). Similarly, the ideas that built the foundations of the HNN are applied to the DE algorithm in this work. This was done to improve the convergence capabilities of the DE algorithm.

First, the population of vectors, x_i^G, was initialized. Then, a set of random weights w_{ij} was introduced and the symmetric property was imposed. The consequence steps are similar to the regular DE algorithm where one principal parent, x_i^p, and three auxiliary parents, x_i^a, are randomly selected. Differential mutation is then performed and the mutated vector, V_i, is generated. The V_i is then recombined with x_i^G to generate child trial vector, x_i^{child}. Consequently, the function A (which arguments are the weights w_{ij} and the child trial vector, x_i^{child}) is introduced as follows:

$$A_i(t) = \sum_{ij} w_{ij}(t) k^{x_i^{child}} \tag{8.13}$$

where k is a constant which is an input parameter specified by the user. This way $A_i(t)$ replaces and changes the functionality of the x_i^{child} in the regular DE algorithm. Then the piece-wise threshold function, $s_i(t)$, is computed as follows:

$$s_i(t+1) = \begin{cases} +1 & \text{iff} \quad A_i(t) > U \\ -1 & \text{iff} \quad A_i(t) < U \\ s_i(t) & \text{iff} \quad \text{otherwise} \end{cases} \tag{8.14}$$

Algorithm 8.2: Hopfield DE (Hopf-DE)

Step 1: Initialize parameters *N*, *P*, *CR*, and *F*
Step 2: Randomly initialize the population vectors, x_i^G
Step 3: Randomly initialize weights, $w_{ij}(t)$
Step 4: Enforce symmetry condition on weights
Step 5: Randomly select one principal parent, x_i^p
Step 6: Randomly select three auxiliary parents, x_i^a
Step 7: Perform differential mutation and generate mutated vector, V_i
Step 8: Recombine V_i with x_i^p to generate child trial vector, x_i^{child}
Step 9: Compute the function $A_i(t)$
Step 10: Compute the energy function, E
Step 11: Perform "knock-out" competition for next-generation survival selection
Step 12: If $dE < 0$, proceed to next step else go to step 4
Step 13: If the fitness criterion is satisfied and $t = T_{max}$, halt and print solutions, else proceed to step 6

---

where U is a constant which is an input parameter specified by the user. The energy function is then computed as follows:

$$E = -\frac{1}{2}\sum_{j=1}\sum_{i=1} A_j(t+1)A_j(t)w_{ij} - \sum_{i=1} \theta s(t) \tag{8.15}$$

where θ is a relaxation constant defined by the user.

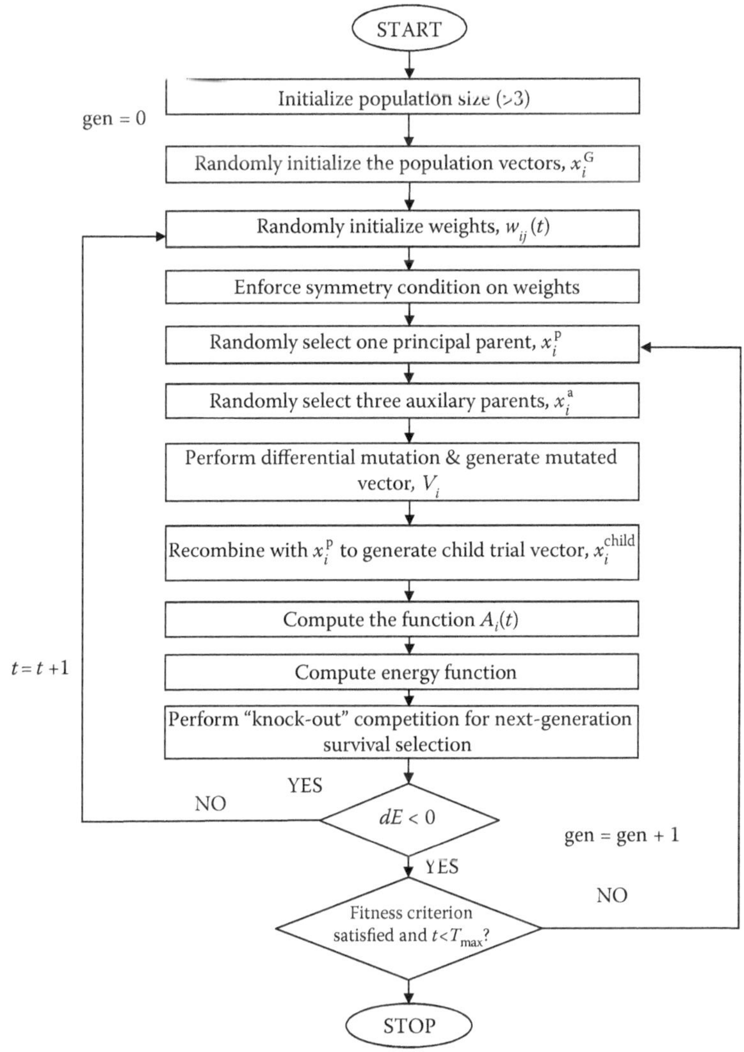

FIGURE 8.4 Hopfield DE (Hopf-DE) flowchart.

Hence, new variant for the DE algorithm is developed and is termed the Hopfield DE (Hopf-DE) algorithm. The workflow of the Hopf-DE algorithm is given in Figure 8.4. The algorithm and the flowchart for the Hopf-DE method are given in Algorithm 8.2.

8.7 DETAILS ON THE COMPUTATIONAL EXPERIMENTS

The weighted sum approach was used in conjunction with the DE method to solve cement-bonded sand mold MO problem. The objective functions (Equations

8.1 and 8.2) were aggregated to form a single weighted function (F) such as the following:

$$F = w_1H + w_2\text{CS} \quad \text{such that} \quad w_i \in [0,1] \quad \text{and} \quad \sum_{i<3} w_i = 1 \tag{8.16}$$

where w_1 and w_2 are the scalar weights. The solution sets (approximations of the Pareto frontier) were obtained using the DE and Hopf-DE techniques in conjunction with the weighted sum approach. The quality of these solutions was measured by using the value of the aggregated function, F. Since this problem involves objective maximization, the higher the F value the more dominant the solution. For the approximation of the Pareto frontier, 50 solutions for various weights were obtained using both the DE and the Hopf-DE methods. The approximate Pareto frontiers traced using the DE algorithm are shown in Figure 8.5.

The maximum, median, and minimum values of the aggregated objective function, F, and their associated weights are given in Table 8.1. Based on these values the best, median, and worst individual solutions were identified. These values are shown in Table 8.2.

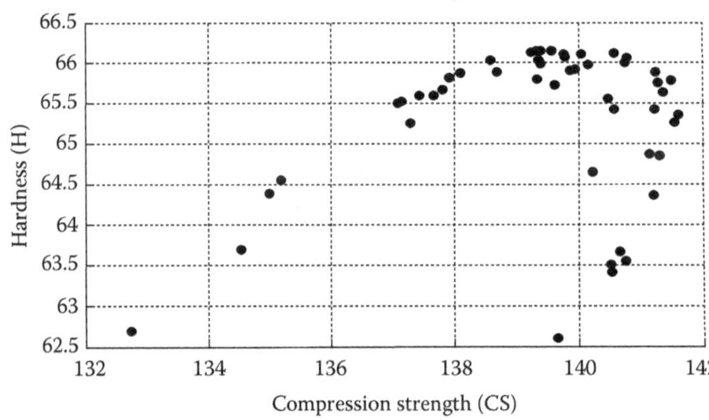

FIGURE 8.5 Approximate Pareto frontiers obtained using the DE algorithm.

TABLE 8.1
Ranked Solutions and the Associated Weights Produced by the DE Algorithm

Description	Best	Median	Worst
F	143.811	101.198	66.8648
w_1	0.99	0.49	0.01
w_2	0.01	0.51	0.99

TABLE 8.2
Ranked Solutions and the Respective Computational Time Produced by the DE Algorithm

Description	Best	Median	Worst
CS	139.537	140.517	139.354
H	66.1537	63.4215	66.1326
A	11.9707	11.9137	11.962
B	3.79641	2.4704	3.82441
C	3.09651	3.05401	3.09731
D	2.09651	2.05401	2.09731
Number of iterations	80	31	59
Computational time (s)	0.748	0.28985	0.55165

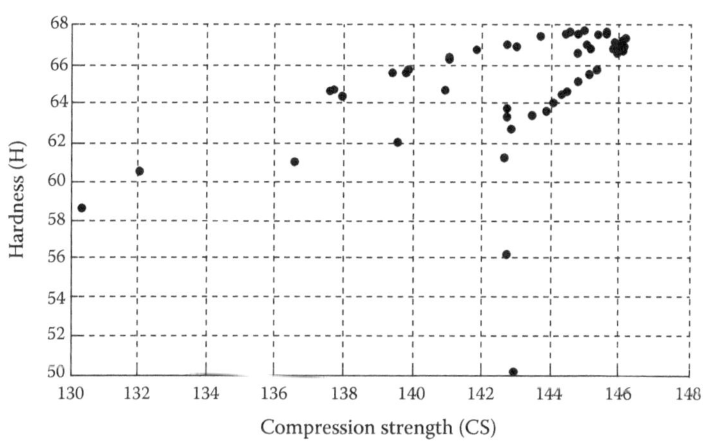

FIGURE 8.6 Approximate Pareto frontiers obtained using the Hopf-DE algorithm.

The approximate Pareto frontiers traced using the Hopf-DE algorithm are shown in Figure 8.6.

Similarly, the solutions produced by the Hopf-DE algorithm were ranked (maximum/best, median, and the minimum/worst) based on the values of the aggregated objective function, *F*. The ranked solutions and their associated weights are given in Table 8.3 while the values of the objective functions, decision variables, and the computational time are provided in Table 8.4.

The total computational time taken by the DE and the Hopf-DE algorithms to obtain the optimal solutions that construct the Pareto frontier is given in Figure 8.7.

The DE algorithm takes 14.92 seconds while the Hop-DE algorithm takes 107.73 seconds to compute the optimal solutions for the entire Pareto frontier. In Figure 8.4, it can be observed that the DE algorithm is more computationally efficient as compared to the Hopf-DE algorithm by approximately 6.175% although in Tables 8.2 and 8.4 it can be seen that Hopf-DE algorithm outperforms the DE algorithm in terms of

TABLE 8.3
Ranked Solutions and the Associated Weights Produced by the Hopf-DE Algorithm

Description	Best	Median	Worst
F	144.591	92.5894	67.0032
w_1	0.99	0.33	0.01
w_2	0.01	0.67	0.99

TABLE 8.4
Ranked Solutions and the Respective Computational Time Produced by the Hopf-DE Algorithm

Description	Best	Median	Worst
CS	145.369	143.686	141.049
H	67.5438	67.4224	66.2552
A	11.9336	11.8749	11.2922
B	3.65916	3.9954	3.94493
C	3.0263	3.03162	3.03082
D	2.01244	2.01225	2.01041
Number of iterations	225	48	119
Computational time (s)	0.91783	0.1958	0.48543

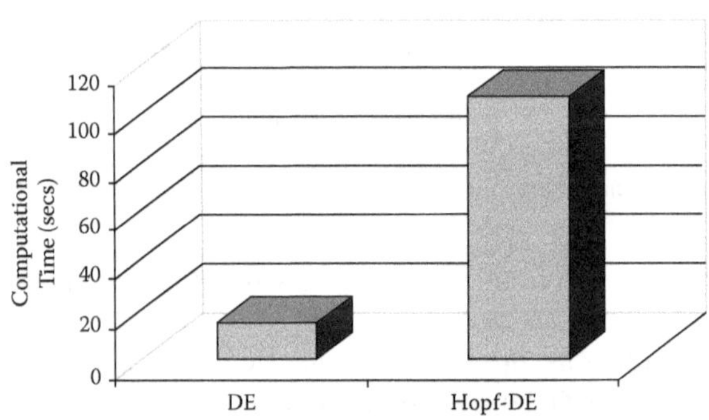

FIGURE 8.7 Computational time for the DE and Hopf-DE techniques.

solution optimality. This may be attributed to the increase in algorithmic complexity of the Hopf-DE technique due to the addition of the Hopfield component in the original DE technique. Therefore, although the Hopf-DE may be superior in terms of obtaining Pareto-efficient solutions, nevertheless, it is computationally inferior as compared to the DE algorithm. The values of the aggregated function, F, produced by the DE and Hopf-DE algorithms with respect to various weights are shown in Figures 8.8 and 8.9 respectively.

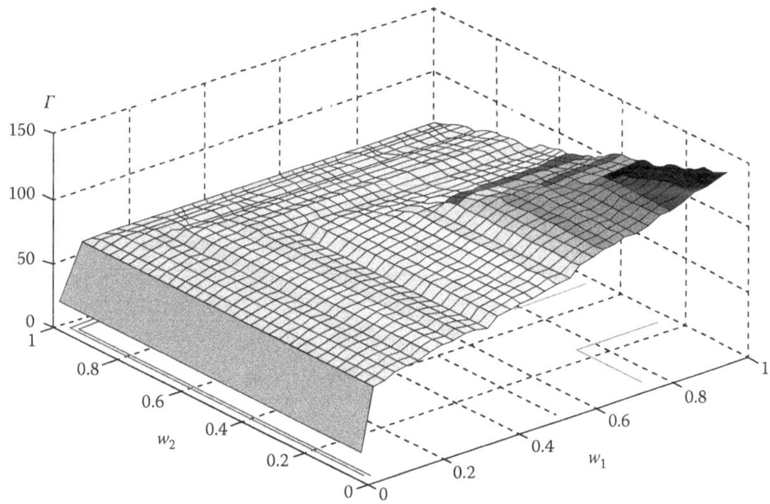

FIGURE 8.8 Values of the aggregated function, *F*, produced by the DE algorithm with respect to the weights.

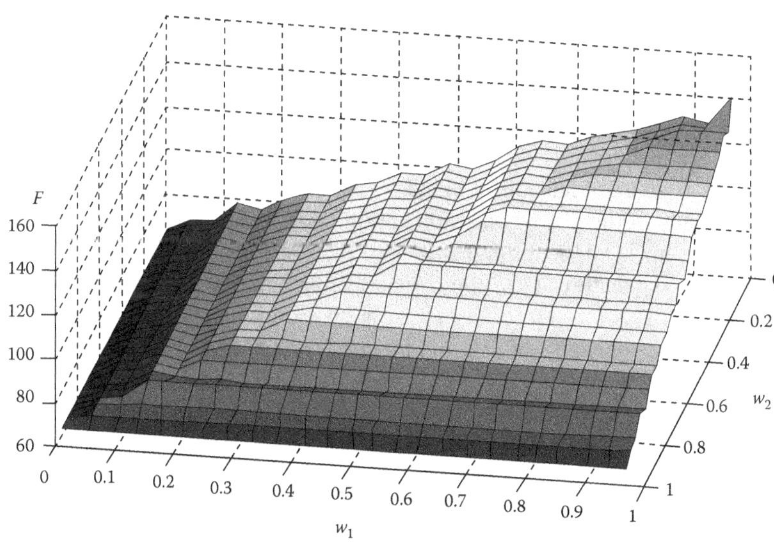

FIGURE 8.9 Values of the aggregated function, *F*, produced by the Hopf-DE algorithm with respect to the weights.

It can be seen from Figures 8.8 and 8.9 that the values of *F* increase sharply as $w_1 \rightarrow 1$ and as $w_2 \rightarrow 0$. In material sciences, it is well known that CS conflicts with hardness (*H*) for most of the known materials. However, in this case, prioritizing hardness (*H*) and neglecting CS boosts up the aggregated function, *F*, value and not vice versa. This characteristic may be mainly attributed to the unique nature of the cement-bonded mold system that the objectives are intercorrelated in such a fashion.

Hence, it can be said that as the second objective (CS) is compromised and the first objective (H) is prioritized, the higher the value of F obtained (optimal value).

From Tables 8.2 and 8.4, it can be stated that Hopf-DE method outperforms the DE technique in terms of obtaining optimal values of CS and hardness (H). This may be attributed to the enhanced capabilities of the Hopfield component in handling nonlinearities since the objective functions used in this problem are nonlinear in nature. With modifications to the energy function of the Hopf-DE as well as its respective parameters, the Hopf-DE method could be customized for solving a broader class of nonlinear problems. This capability can be mainly attributed to the convergence capabilities of the Hopfield component to handle factors arising from problems which are multivariate and nonlinear.

The DE and the Hopf-DE methods make on average 31.92 and 524.96 function evaluations per optimal solution point on the Pareto frontier. Thus, the DE method is the most efficient in program execution as compared to the Hopf-DE method. Similar to computational time the prime contributing factor to this difference is algorithmic complexity where the Hopf-DE method has a higher degree of complexity as compared to the DE method. Due to the energy function component in the Hopf-DE method, the algorithm showed stable performance and converged during each program execution.

A global solution with a better computational time may be obtained by implementing the DE and Hopf-DE methods on a high-performance computer (HPC). The HPC should be a dedicated system capable of rigorous search and processing. It is crucial to remark that the Hopf-DE and DE methods were easily implemented using the language of C++. During the implementation of the DE and the Hopf-DE algorithms on the cement-bonded mold application, both methods performed very well in terms of solution feasibility. Hence, solutions produced by all three methods did not break any of the given constraints in this problem.

An increase in the algorithmic features to include more optimization as well as constraint handling features may improve both approaches. Hence, this strategy may produce solution sets that are closer to the global optimal. Since the execution of these sorts of programs (algorithmically complex) would be computationally costly, a more powerful computer would be required. Another method to improve the three methods utilized in this work would be to further hybridize it with other methods like simulator annealing (Kirkpatrick et al., 1983), genetic programming (Koza, 1992), or other classes of evolutionary algorithms. Hybridizing these methods with algorithms like DE may provide it with a more efficient system for handling constraints and thus pave the way for a solution closer to the global optimum.

8.8 SUMMARY

The main ideas presented in this chapter involve the implementation of the DE and the improved Hopf-DE techniques in material engineering. The DE and Hopf-DE approaches were applied to the MO cement-bonded sand mold system to construct the approximate Pareto frontier which was then employed to identify the best solution option. Prior to the implementation of the DE and Hopf-DE techniques, other metaheuristics such as swarm-based algorithms were attempted on the cement-bonded sand mold problem. However, due to poor results these approaches were abandoned and the DE technique was instead applied to this problem. Due to the "No Free Lunch Theorems" (Wolpert & Macready, 1997), not all problems can be tackled using the same metaheuristic strategy. This fact applies to any form of computational technique. Thus, a certain degree of algorithmic testing is required prior to the identification of a suitable metaheuristic for a particular problem. Detailed analyses were performed on the computational results produced by the DE and its variant the Hopf-DE. Examination of the performance and the quality of the solutions obtained by the DE and Hopf-DE algorithms was presented.

REFERENCES

Ali, M., Pant, M., & Abraham, A. (2009). Simplex differential evolution. *Acta Polytechnica Hungarica*, *6*(5), 95–115.

Amit, D. J., Gutfreund, H., & Sompolinsky, H. (1986). Spin-glass models of neural networks. *Physical Review A*, *32*(2), 1007–1018.

Anderson, M. J., & Whitcomb, P. J. (2005). *RSM simplified: Optimizing processes using response surface methods for design of experiments*. New York: Productivity Press.

Babu, B. V., & Angir, R. (2003). New strategies of differential evolution for optimization of extraction process. *Proceedings of International Symposium & 56th Annual Session of IIChE (CHEMCON-2003)*, Bhubaneswar, India.

Bhattacharya, A. K., Debjani, S., Roychowdhury, A., & Das, J. (2007). Optimization of continuous casting mold oscillation parameters in steel manufacturing process using genetic algorithms. *IEEE Congress on Evolutionary Computation* (pp. 3998–4004), Piscataway, NJ: IEEE.

Bhattacharya, A. K., & Sambasivam, D. (2009). *Optimization of oscillation parameters in continuous casting process of steel manufacturing: Genetic algorithms versus differential evolution*. INTECH Open Access Publisher.

Das, S., Abraham, A., & Konar, A. (2008). Particle swarm optimization and differential evolution algorithms: Technical analysis, applications and hybridization perspectives. *Studies in Computational Intelligence (SCI)*, 1–38.

Dyson, F. J. (1969). Existence of a phase-transition in a one-dimensional Ising ferromagnet. *Communications in Mathematical Physics*, *12*, 91–107.

Elamvazuthi, I., Ganesan, T., & Vasant, P. (2011). A comparative study of HNN and hybrid HNN-PSO techniques in the optimization of distributed generation (DG) power systems. *2011 International Conference on Advanced Computer Science and Information System (ICACSIS) IEEE* (pp. 195–200), Piscataway, NJ: IEEE.

Fishburn, P. C. (1967). Additive utilities with incomplete product set: Applications to priorities and assignments. *Operations Research*, *15*, 537–542.

Ganesan, T., Elamvazuthi, I., Shaari, K. Z. K., & Vasant, P. (2013). Swarm intelligence and gravitational search algorithm for multi-objective optimization of synthesis gas production. *Journal of Applied Energy*, *103*, 368–374.

Ganesan, T., Vasant, P., & Elamvazuthi, I. (2011). Evolutionary normal-boundary intersection (ENBI) method for multi-objective optimization of green sand mold system, control system, computing and engineering (ICCSCE), *2011 IEEE International Conference,* November 25–27, 2011 (pp. 86–91), Penang, Malaysia.

Ganesan, T., Vasant, P., Elamvazuthi, I., & Shaari, K. Z. K. (2012). Multiobjective optimization of green sand mold system using DE and GSA. *2012 4th International Conference of Soft Computing and Pattern Recognition* (SoCPaR), December 10–13, 2012 (pp. 255–259), Brunei.

Guharaja, S., Noorul Haq, A., & Karuppannan, K. (2006). Optimization of green sand casting process parameters by using Taguchi's method. *The International Journal of Advanced Manufacturing Technology*, *30*(11–12), 1040–1048.

Holis, M., Skanderova, L., Placek, M., Dvorsky, J., & Zelinka, I. (2013). Solving steel alloying using differential evolution and SOMA. *Computer Information Systems and Industrial Management.* (Lecture Notes in Computer Science Vol. 8104, pp. 453–464). Berlin: Springer.

Holland, J. H. (1992). Adaptation in natural and artificial systems: An introductory analysis with applications to biology. *Control and artificial intelligence.* Cambridge, MA: MIT Press.

Hopfield, J. J. (1982). Neural networks and physical systems with emergent collective computational abilities. *Proceedings of National Academy of Sciences USA*, *29*, 2554–2558.

Hopfield, J. J. (1984). Neurons with graded response have collective computational properties like those of two-state neurons. *Proceedings of National Academy of Sciences USA*, *81*, 3088–3092.

Ingber, L. (1993). Simulated annealing: Practice versus theory. *Mathematical and Computer Modelling*, *18*, 29–57.

Jawaha, S., & Ramamoorthy, P. (2012). Dynamic optimization of injection molding process variables by evolutionary programming methods. *International Conference on Computer Communication and Informatics (ICCCI–2012)*, January 10–12, 2012 (pp. 690–695), Coimbatore, India.

Ji, Z., & Jiang, Y. (2012). Optimization of cooling parameters in casting processes based on adaptive evolutionary particle swarm algorithm. *Advanced Materials Research*, *468–471*, 2399–2402.

Keswani, I. P., & Upadhye, R. A. (2012). Optimization of sand casting process parameter using Taguchi method in foundry. *International Journal of Engineering Research & Technology (IJERT)*, *1*(7), 1–11.

Kim, B. H., Shin, J. S., Lee, S. M., & Moon, B. M. (2007). Improvement of tensile strength and corrosion resistance of high-silicon cast irons by optimizing casting process parameters. *Journal of Material Science*, *42*, 109–117.

Kirkpatrick, S., Gelatt, C. D., & Vecchi, M. P. (1983). Optimization by simulated annealing. *Science*, *220*, 671–680.

Koza, J. R. (1992). *Genetic programming: On the programming of computers by means of natural selection.* Cambridge, MA: MIT Press.

Kumar, A., & Singh, J. (2011). Optimization of sand casting process parameters for ferrous material by using design of experiments. *International Journal on Emerging Technologies*, *2*(2), 122–127.

Lagdive, P. B., & Inamdar, K. H. (2013). Optimization of riser in casting using genetic algorithm. *International Archive of Applied Sciences and Technology*, *4*, 21–26.

Lee, K. Y., Sode-Yome, A., & Park, J. H. (1998). Adaptive Hopfield neural networks for economic load dispatch. *IEEE Transactions on Power Systems*, *13*(2), 519–526.

Metropolis, N., Rosenbluth, A. W., Rosenbluth, M., Teller, A. H., & Teller, E. (1953). Equation of state calculations by fast computing machines. *Journal of Chemical Physics*, *21*, 1087–1092.

Montgomery, D.C. (2001). *Design and analysis of experiments* (5th ed.). New York: John Wiley & Sons.

Neville, A. M. (2000). *Properties of concrete* (4th ed.). Essex, England: Pearson Education Limited.

Parappagoudar, M. B., Pratihar, D. K., & Datta, G. L. (2007). Non-linear modeling using central composite design to predict green sand mold properties. *Proceedings IMechE B Journal of Engineering Manufacture*, *221*, 881–894.

Sleicher, C. A., & Providence, R. I. (1943). Use of cement in foundry molding. *AFS Transactions*, *51*, 737–747.

Storn, R., & Price, K. V. (1995). Differential evolution—A simple and efficient adaptive scheme for global optimization over continuous spaces, ICSI, Technical Report TR-95–012.

Surekha, B., Vudavilli, P. R., & Parappagoudar, M. B. (2012). Modelling of cement-bonded sand mold system: An artificial intelligence approach. *International Journal of Manufacturing, Materials and Mechanical Engineering*, *2*(1), 30–46.

Tank, D. W., & Hopfield, J. J. (1986). Simple neural optimization network: An A\D converter, signal decision circuit and a linear programming circuit. *IEEE Transactions on Circuits and Systems*, *CAS-33*, 533–541.

Triantaphyllou, E. (2000). *Multi-criteria decision making: A comparative study* (p. 320). Dordrecht, The Netherlands: Kluwer Academic Publishers (now Springer).

Uchikawa, H., & Uchida, S. (1976). On the special super-high early-strength cement sand process. AFS *International Cast Metals Journal, 1*, 3–31.

Vasant, P., Ganesan, T., & Elamvazuthi, I. (2012). Hybrid tabu search hopfield recurrent ANN fuzzy technique to the production planning problems: A case study of crude oil in refinery industry. *International Journal of Manufacturing, Materials, and Mechanical Engineering*, *2*(1), 47–65.

Wolpert, D. H., & Macready, W. G. (1997). No free lunch theorems for optimization. *IEEE Transactions on Evolutionary Computation*, *1*, 67.

Xudong, W., Zhaofeng, W., & Man, Y. (2011). Particle swarm optimization for estimation on mold oscillation for continuous casting process. *Seventh International Conference on Natural Computation*, July 26–28, 2011 (pp. 2226–2229), Donghua University, Shanghai, China.

9 Multiobjective Optimization of Green Sand Mold Product

Various improvements have been made by researchers, engineers, and decision makers in the foundry industry to obtain high-quality products at reasonable operational costs and minimal energy consumption. Thus, various optimization approaches have been practiced by personnel in this industry ranging from experiential strategies to statistical and metaheuristic-based approaches. The primary recipe for achieving the mentioned goals in the foundry industry is by the expert manipulation of the parameters involved in the molding process.

9.1 MULTIOBJECTIVE OPTIMIZATION SCENARIO

Many concerns have been raised when dealing with emerging technologies in engineering optimization which present themselves in a multiobjective (MO) setting (Eschenauer, Koski, & Osyczka, 1990; Statnikov & Matusov, 1995). Strategies in MO optimization can be rudimentarily classified into two groups. First, there are methods that use the concept of Pareto optimality to trace the non-dominated solutions at the Pareto curve, for instance: strength Pareto evolutionary algorithm (Zitzler & Thiele, 1998) and non-dominated sorting genetic algorithm II (NSGA-II) by Deb, Pratap, Agarwal, and Meyarivan (2002). The second class of techniques is known as the weighted (or scalarization) approaches. During the application of these methods, the objective functions are aggregated into a single weighted function which is then solved for various scalar (weight) values. Some well-known scalarization techniques include the weighted sum method (Triantaphyllou, 2000), goal programming (Luyben & Floudas, 1994), and normal-boundary intersection (NBI) method (Das & Dennis, 1998). Using these techniques, the scalars (or weights) are used to consign relative preferences to the objectives during the aggregation procedure. Hence, alternative near-optimal solution options are generated for various values of the scalars.

In MO optimization problems, determining the most efficient solution set can be a very daunting process. Many varieties of concepts (such as diversity and convergence) have been proposed in the past years. These ideas were then used as indicators to evaluate the solution set produced by the optimization algorithm (Grosan, 2003). Such evaluations were then used to benchmark the algorithm's performance. These concepts unfortunately could not absolutely state and rank the superiority of solution sets produced by an algorithm against other such sets generated by other algorithms. The only concept that can be used generally for the overall ranking of solution sets is the idea of "Pareto-dominance." The hypervolume indicator (HVI) (Zitzler & Thiele, 1998) is a set measure reflecting the volume enclosed by a Pareto frontier approximation and a reference set (Knowles & Corne, 2003; Igel, Hansen, & Roth, 2007). The HVI thus guarantees strict monotonicity regarding Pareto dominance (Fleischer, 2003; Zitzler, Thiele, Laumanns, Fonseca, & Grunert da Fonseca, 2003). This makes the ranking of solution sets and hence algorithms possible for an arbitrary MO problem.

This chapter discusses the procedures and outcomes involving the implementation of metaheuristics to dominantly approximate the Pareto frontier in the objective space. Three non-modified techniques which are differential evolution (DE), particle swarm optimization (PSO), and gravitational search algorithm (GSA) are implemented on the green sand system. In addition, the modified versions of PSO and DE techniques which are chaotic DE (CDE), chaotic PSO, and game-theoretic DE approaches are developed and applied to the mentioned problem. These techniques are utilized in conjunction with the weighted sum method to improve the dominance of the solution set during successive iterations. In addition, the solution characteristics (diversity and degree of dominance) of the solutions with regard to their level of dominance are also measured using two performance metrics. The next section provides details on the green sand mold system and its associated process model.

9.2 GREEN SAND MOLD SYSTEM

In green sand mold systems, the quality of the product obtained from the molding process is very much dependent on the physical properties of the molding sand (e.g., hardness, permeability, green compression strength, and bulk density). Faulty extent of the mentioned properties may result in casting defects such as poor surface finish, blowholes, scabs, pinhole porosity, and so on. Controllable parameters such as percentage of water, percentage of clay, grain fineness number, and number of strokes heavily influence the physical properties of the molded sand. Therefore, by classifying these parameters as decision variables and mold sand properties as the objective function, the MO optimization problem was formulated in Surekha, Kaushik, Panduy, Vundavilli, and Parappagoudar (2012). The purpose of this formulation is the determination of the best controllable parameters for optimal final product of the molding process. A more rigorous study on the optimization and model development of mold systems can be seen in Sushil, Satsangi, and Prajapati (2010).

In the green sand mold system, the response parameters of the mold heavily influence the quality of the final product. In Surekha et al. (2012), these parameters are selected as the objective functions. The response parameters are: green compression strength (f_1), permeability (f_2), hardness (f_3), and bulk density (f_4). These objectives on the other hand are influenced by the process variables which are: the grain fineness number (A), percentage of clay content (B), percentage of water content (C), and number of strokes (D). The objective functions and the range of the decision variables are shown as follows:

$$\begin{aligned} f_1 &= 17.2527 - 1.7384A - 2.7463B + 32.3203C + 6.575D \\ &+ 0.014A^2 + 0.0945B^2 - 7.7857C^2 - 1.2079D^2 \\ &+ 0.0468AB - 0.1215AC - 0.0451AD \\ &+ 0.5516BC + 0.6378BD + 2.689CD \end{aligned} \tag{9.1}$$

$$\begin{aligned} f_2 &= 1192.51 - 15.98A - 35.66B + 9.51C - 105.66D \\ &+ 0.07A^2 + 0.45B^2 - 4.13C^2 + 4.22D^2 + 0.11AB \\ &+ 0.2AC + 0.52AD + 1.19BC + 1.99BD - 3.1CD \end{aligned} \tag{9.2}$$

$$\begin{aligned} f_3 &= 38.2843 - 0.0494A + 2.4746B + 7.8434C + 7.774D \\ &+ 0.001A^2 - 0.00389B^2 - 1.6988C^2 - 0.6556D^2 \\ &- 0.0015AB - 0.0151AC - 0.0006AD \\ &- 0.075BC - 0.1938BD + 0.65CD \end{aligned} \tag{9.3}$$

$$\begin{aligned} f_4 &= 1.02616 + 0.01316A - 0.00052B - 0.06845C \\ &+ 0.0083D - 0.00008A^2 + 0.0009B^2 + 0.0239C^2 \\ &- 0.00107D^2 - 0.00004AB - 0.00018AC + 0.00029AD \\ &- 0.00302BC - 0.00019BD - 0.00186CD \end{aligned} \tag{9.4}$$

$$\begin{aligned} &52 \le A \le 94 \\ &8 \le B \le 12 \\ &1.5 \le C \le 3 \\ &3 \le D \le 5 \end{aligned} \tag{9.5}$$

To obtain the size distributions of the silica sand and the grain fineness number, sieve analysis tests were carried out in Parappagoudar, Pratihar, and Datta (2007). Similarly, in Surekha et al. (2012), the authors conducted gelling index tests for the determination of the strength of clay. Next, experiments were conducted by varying the combination of the parameters using the central composite design. The mathematical model of the green mold system was developed; its objective functions are given in Equations 9.1 through 9.4 and the constraints are given in Equation 9.5. The MO optimization problem statement for the green mold system problem is shown as follows:

$$\begin{aligned}&\text{Max} \quad (f_1, f_2, f_3, f_4) \quad \text{subject to}\\&52 \le A \le 94\\&8 \le B \le 12\\&1.5 \le C \le 3\\&3 \le D \le 5\end{aligned} \tag{9.6}$$

9.3 PERFORMANCE METRICS

In the past two decades, visual observation is the most common method employed to gauge the degree of diversity and convergence to the Pareto frontier (Mostaghim & Teich, 2003). For cases involving low-dimensional objective spaces ($n < 3$) these methods are deemed sufficient. However, for cases involving high-dimensional objective spaces ($n \geq 3$), analysis by visualization per se is very difficult to practice and is a very inaccurate method to draw any information out of the solution vectors in the objective space. Thus, many quantitative performance metrics have been proposed and used effectively in recent times. A concise and complete review on the developments of performance metrics were given in Okabe, Jin, and Sendho (2003). The application of the convergence and diversity metrics to test problems was done in Knowles and Corne (2003) and Farhang-Mehr and Azarm (2002) while the HVI was utilized for dominance determination in Zitzler (1999). The convergence and diversity metrics used in this work are specifically for cases involving high-dimensional objective spaces. In this work, real-world industrial cases with high-dimensional objective spaces are solved. These cases provide no information about the Pareto-optimal front. Therefore, the performance metrics essentially deliver convergence and diversity information relative to an approximated set.

9.3.1 Sigma Diversity

The sigma diversity metric (SDM) evaluates the locations of the solution vectors in the objective space relative to the sigma vectors (Mostaghim & Teich, 2003). For lower-dimensional objective spaces ($n < 3$), metrics which are based on spherical and polar coordinates may be used. However, as the dimensions increase beyond three ($n \geq 3$), the mentioned coordinate systems do not define the distribution of the solution vectors as well. In such scenarios, the SDM is highly recommended for effective computing of the solution distribution.

To begin the computation of the SDM, two types of sigma lines would have to be constructed. There are two types of sigma lines: those that represent the solution vectors (σ') and lines that represent the reference lines (σ). The sigma lines that represent the solution vectors can be computed as follows:

$$\sigma'_k(ij) = \frac{f_j^2 - f_j^2}{\sum_{l=1}^{n} f_l^2} \quad \text{such that} \quad \forall i \neq j \tag{9.7}$$

where k denotes the index that represents the number of solution vectors; i, j, and l denote the index that represents the number of objectives; and n denotes the total number of objectives. Then the magnitude sigma $|\sigma'_k|$ is computed as follows:

$$|\sigma'_k| = \sqrt{\sum_{i=1}^{m} \sum_{j=1}^{m} \sigma'_k(ij)} \tag{9.8}$$

Thus, for each line in the objective space (solution vector or reference line), there exists a unique sigma value. For the sake of clarity, a two-objective distribution of the reference sigma line as well as the solution vector sigma line is given in Figure 9.1.

Next is the task to determine the reference sigma lines, $|\sigma|$. It is crucial to take note that the total number of reference sigma lines must be equivalent to the total number of solution vector sigma lines which is also equivalent to the number of solution vectors present. For instance, let there be a total of k amount of $|\sigma|$ lines, p amount of $|\sigma'|$ lines, and m be the total number of solution vectors in some arbitrary n-dimensional objective space. Then, $k = l = m$. The reference sigma lines have to be uniformly distributed in the objective space. To achieve this, the most minimum and the maximum values of $|\sigma'|$ are determined. Then the step interval, Δ, for even distribution is computed as follows:

$$\Delta = \frac{\max(|\sigma'|) - \min(|\sigma'|)}{m} \tag{9.9}$$

Next, the reference sigma lines, $|\sigma|$, are determined as follows:

$$|\sigma_{k+1}| = |\sigma_k| + \Delta \quad \text{such that} \quad |\sigma_0| = \min(|\sigma'_k|) \tag{9.10}$$

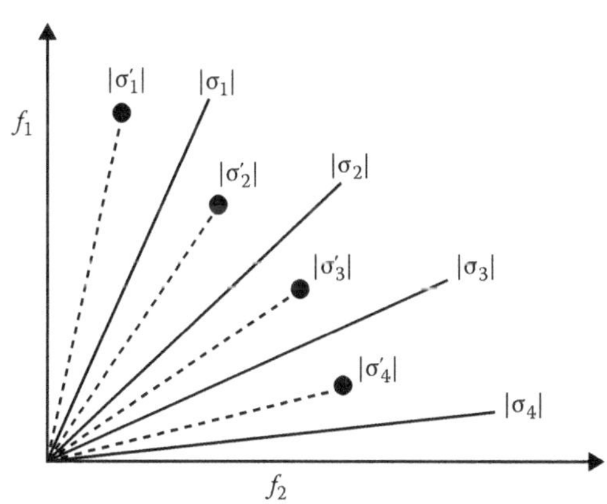

FIGURE 9.1 Two-dimensional visualization of the solution vectors and the sigma lines.

To compute the metric, the distance of the solution vector sigma lines $|\sigma'|$ from the reference sigma lines $|\sigma|$, θ_k, is calculated by taking the difference of its respective values such as

$$\theta_k = |\sigma'_k| - |\sigma_k| \tag{9.11}$$

The piece-wise satisfaction condition, μ_k, is then imposed such that it returns binary values relative to some user-specified threshold distance, c, as follows:

$$\mu_k = \begin{cases} 1 & \text{iff} \quad \theta \leq c \\ 0 & \text{iff} \quad \theta > c \end{cases} \tag{9.12}$$

The sigma diversity metric, $D_{k,}$ is then calculated by averaging the piece-wise satisfaction condition, μ_k, over the number of solution vectors, m:

$$D_k = \frac{\sum_{k=1}^{m} \mu_k}{m} \tag{9.13}$$

The algorithm for computing the sigma diversity metric, D_k is shown in Algorithm 9.1.

Alogrithm 9.1: Sigma Diversity Computation

Step 1: Determine the sigma lines for the solution vectors, $|\sigma'_k|$
Step 2: Compute the magnitude sigma $|\sigma'_k|$ for each of the solution vectors
Step 3: Calculate the step interval, Δ
Step 4: Construct the reference sigma lines $|\sigma_k|$
Step 5: Compute the distance, θ_k, of sigma lines $|\sigma'|$ from $|\sigma|$
Step 6: Impose the piece-wise satisfaction condition and compute μ_k
Step 7: Compute sigma diversity metric, D_k

Thus, high values of the sigma diversity metric, D_k, indicates high uniformity and diversity in terms of the distributions of the solution vectors in the objective space.

9.3.2 Hypervolume Indicator

Many past investigations have been conducted on methods to apply evolutionary algorithms to MO problems. In addition, many ideas have also been proposed on methods to attain a good solution spread over the Pareto frontier. Unfortunately, in all these works, the optimization goal remains vague since there are no established techniques that measure the quality of a solution sets produced by an arbitrary MO algorithm (especially in cases where there are more than two objectives in the problem). The impact of this issue magnifies especially in real-world problems where it is often the case that the Pareto frontier is unknown (unlike theoretical test functions where the Pareto frontier is known and can be used to benchmark the solution quality).

One approach that has been effective in measuring the quality of the solution set that constructs the Pareto frontier in cases where the Pareto frontier is unknown is the HVI (Beume, Naujoks, & Emmerich, 2007). Recently, this indicator has been frequently applied in many works involving MO problems (Igel et al., 2007). The HVI is the only indicator which is strictly Pareto compliant that can be used to measure the quality of solution sets in MO optimization problems (Zitzler & Thiele, 1998). To provide a clearer picture of the idea of Pareto compliance, the concepts of Pareto dominance are formally defined in the subsequent paragraphs.

Pareto dominance can be categorized into three types which are strictly dominates ($\succ$), weakly dominates ($\succeq$), and indifferent ($\sim$). Let two solution vectors be **a** and **b**. Then if the solution vector **a** dominates the vector **b** in all the objectives then **a** strictly dominates **b** (**a** $\succ$ **b**). If the solution vector **a** dominates the vector **b** in some of the objectives but not all, then **a** weakly dominates **b** (**a**$\succeq$**b**). Finally, if the solution vector **a** does not dominate the vector **b** and the solution vector **b** does not dominate **a** as well in all the objectives, then **a** is indifferent to **b** (**a** $\sim$ **b**).

Strictly Pareto compliance can be defined as follows. Let there be two solution sets say X and Y for a particular MO problem. If the HVI value for X is greater than that of Y, then the solution set $X \succ Y$ or $X \succeq Y$. The HVI measures the volume of the dominated section of the objective space and can be applied for multidimensional scenarios. When using the HVI, a reference point or a "nadir point" needs to be defined. The nadir point is a point which is dominated by all the solutions from the approximate Pareto frontier. Relative to this point, the volume of the space of all dominated solutions can be computed. The HVI of a solution set $x_d \in X$ can be defined as follows:

$$\mathrm{HVI}(X) = \mathrm{vol}\left(\bigcup_{(x_1,\ldots,x_d)\in X} [r_1, x_1] \times \cdots \times [r_d, x_d] \right) \tag{9.14}$$

where $r_1,\ldots,r_d$ is the reference point and vol(.) is the Lebesgue measure. For illustration purposes, a two-objective depiction of the construction of the hypervolume is given in Figure 9.2.

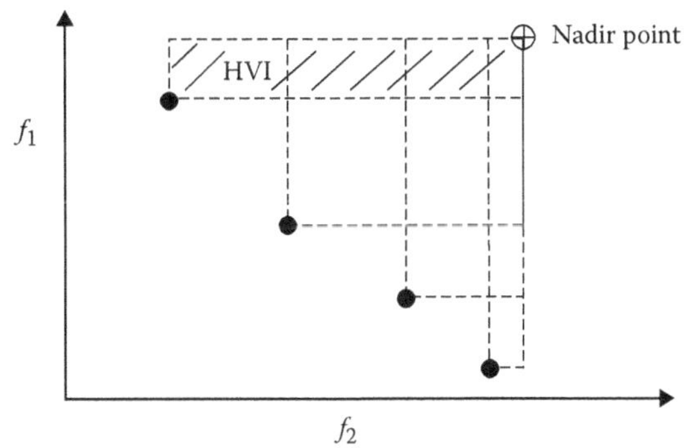

FIGURE 9.2 Hypervolume construction for a two-objective minimization problem.

It can be observed in Figure 9.2 that the larger the value of the HVI, the more dominant the solution is in the objective space. The characteristics of the HVI are as follows:

a. Strictly monotonic indicator.
b. Its computational effort is exponential to the amount of solution vectors.
c. It requires an upper-bounding vector (nadir point).

9.4 GRAVITATIONAL SEARCH ALGORITHM

The GSA algorithm is a metaheuristic algorithm first developed by Rashedi, Nezamabadi-pour, and Saryazdi (2009). This technique was inspired by the law of gravity and the idea of interaction of masses. This algorithm uses the Newtonian gravitational laws where the search agents are the associated masses. Thus, the gravitational forces influence the motion of these masses, where lighter masses gravitate toward the heavier masses (which signifies good solutions) during these interactions. The gravitational force hence acts as the communication mechanism for the masses (analogous to "pheromone deposition" for ant agents in ant colony optimization [Colorni, Dorigo, & Maniezzo, 1991] and the "social component" for the particle agents in PSO). The position of the masses correlates with the solution space in the search domain while the masses characterize the fitness space. As the iterations increase, and gravitational interactions occur, it is expected that the masses would conglomerate at their fittest position and provide an optimal solution to the problem.

Initially the GSA algorithm randomly generates a distribution of masses, $m_i(t)$ (search agents), and also sets an initial position for these masses, x_i^d. For a minimization problem, the least fit mass, $m_i^{\text{worst}}(t)$, and the fittest mass, $m_i^{\text{best}}(t)$, at time t are calculated as follows:

$$m^{\text{best}}(t) = \min_{j \in [1,N]} m_j(t) \tag{9.15}$$

$$m^{\text{worst}}(t) = \max_{j \in [1,N]} m_j(t) \tag{9.16}$$

For a maximization problem, it is simply vice versa. The inertial mass, $m_i'(t)$, and gravitational masses, $M_i(t)$, are then computed based on the fitness map developed previously.

$$m_i'(t) = \frac{m_i(t) - m^{\text{worst}}(t)}{m^{\text{best}}(t) - m^{\text{worst}}(t)} \tag{9.17}$$

$$M_i(t) = \frac{m_i(t)}{\sum_{j=1}^{N} m_j(t)} \tag{9.18}$$

such that

$$M_{ai} = M_{pi} = M_{ii} = M_i : \quad i \in [1, N] \tag{9.19}$$

Then the gravitational constant, $G(t+1)$, and the Euclidean distance, $R_{ij}(t)$, are computed as follows:

$$G(t+1) = G(t)\exp\left(\frac{-\alpha t}{T_{\max}}\right) \tag{9.20}$$

$$R_{ij}(t) = \sqrt{\left(x_i(t)\right)^2 - \left(x_j(t)\right)^2} \tag{9.21}$$

where α is some arbitrary constant, $T_{\max}$ is the maximum number of iterations, and $x_i(t)$ and $x_j(t)$ are the positions of particles i and j at time t. The interaction forces at time t, $F_{ij}^{d}(t)$, for each of the masses are then computed:

$$F_{ij}^{d}(t) = G(t)\left(\frac{M_{pi}(t) \times M_{aj}(t)}{R_{ij}(t) + \varepsilon}\right) \times \left(x_j^{d}(t) - x_i^{d}(t)\right) \tag{9.22}$$

where ε is some small parameter. The total force acting on each mass i is given in a stochastic form as follows:

$$F_i^{d}(t) = \sum_{\substack{j=1 \\ i \neq j}}^{N} \text{rand}(w_j) F_{ij}^{d}(t) :$$
$$\text{rand}(w_j) \in [0,1] \tag{9.23}$$

where $\text{rand}(w_j)$ is a randomly assigned weight. Consequently, the acceleration of each of the masses, $a_i^{d}(t)$, is given as follows:

$$a_i^{d}(t) = \left(\frac{F_i^{d}(t)}{M_{ii}(t)}\right) \tag{9.24}$$

After the computation of the particle acceleration, the particle position and velocity are calculated:

$$v_i^{d}(t+1) = \text{rand}(w_j) + v_i^{d}(t) + a_i^{d}(t) \tag{9.25}$$

$$x_i^{d}(t+1) = x_i^{d}(t(t) + v_i^{d}(t(t+1) \tag{9.26}$$

where $\text{rand}(w_j)$ is a randomly assigned weight. The iterations are then continued until all mass agents are at their fittest positions in the fitness landscape and some stopping criterion which is set by the user is met. The GSA algorithm is presented in Algorithm 9.2.

Algorithm 9.2: Gravitational Search Algorithm (GSA)

Step 1: Initialize number of particles, $m_{i,}$ and initial positions, $x_i(0)$
Step 2: Initialize algorithm parameters, $G(0)$, α
Step 3: Compute gravitational and inertial masses based on the fitness map
Step 4: Compute the gravitational constant, $G(t)$
Step 5: Compute distance between agents, $R_{ij}(t)$
Step 6: Compute total force, $F_i^{\mathrm{d}}(t)$ and the acceleration $a_i^{\mathrm{d}}(t)$ of each agent
Step 7: Compute new velocity $v_i(t)$ and position $x_i(t)$ for each agent
Step 8: If the fitness criterion is satisfied and $t = T_{\max}$, halt and print solutions, else proceed to step 3

9.5 CHAOTIC PSO

The algorithmic enhancement presented here involves the concept of the application of chaotic logistic maps (Lorenz, 1963; Jakobson, 1981; Flake, 1998) to the PSO algorithm to improve its diversification capabilities. Commonly, in most PSO algorithms, a random number generator is used to establish the initial position and velocity distributions of the particles. This would give the initial diversity in the population of solutions represented by each of the particles. The diversification capabilities of the PSO are enhanced by plugging in the chaotic logistic map after the random number generator segment in the algorithm. This way, the random number generator produces the initial conditions, $N_i(0)$, for the chaotic logistic map. The logistic mapping is given as follows:

$$R_i(t) = \lambda N_i(t) \tag{9.27}$$

$$N_i(t+1) = R_i(t)N_i(t)\left[1 - N_i(t)\right] \tag{9.28}$$

$$R_i(t+1) = R_i(t) + \lambda' \tag{9.29}$$

where $N(t)$ and $R(t)$ are variables in the logistic chaotic map , and λ' and λ are relaxation constants specified by the user. The chaotic logistic function is then iterated until the limit $N_{\max}$ is reached. The solutions obtained at the end of these iterations are then used as the initial distributions for the particle position and velocity ($x_i(0)$ and $v_i(0)$). Hence, another variant for the PSO algorithm (chaotic PSO or Ch-PSO) with magnified diversification capabilities is described in this chapter. The algorithm and the flowchart for this Ch-PSO method are given in Algorithm 9.3 and Figure 9.3 respectively.

Algorithm 9.3: Chaotic Particle Swarm Optimization (Ch-PSO)

Step 1: Set number of particles, i, and initialize parameter settings w, c_1, c_2, r_1, r_2, n_0
Step 2: Randomly initialize particles' position $x_i(t)$ and velocity $v_i(t)$
Step 3: Randomly set initial conditions for the chaotic logistic map
Step 4: Iterate chaotic logistic map
Step 5: If the iterations, $n > N_{max}$, proceed to step 6 else go to step 4
Step 6: Calculate cognitive and social components of the particles
Step 7: Compute position $x_i(t+1)$ and velocity $v_i(t+1)$ of the particles at next iteration
Step 8: Proceed with the evaluation of the fitness of each particle in the swarm
Step 9: If the fitness conditions are satisfied and $t < T_{max}$, stop program and print solutions, else go to step 6

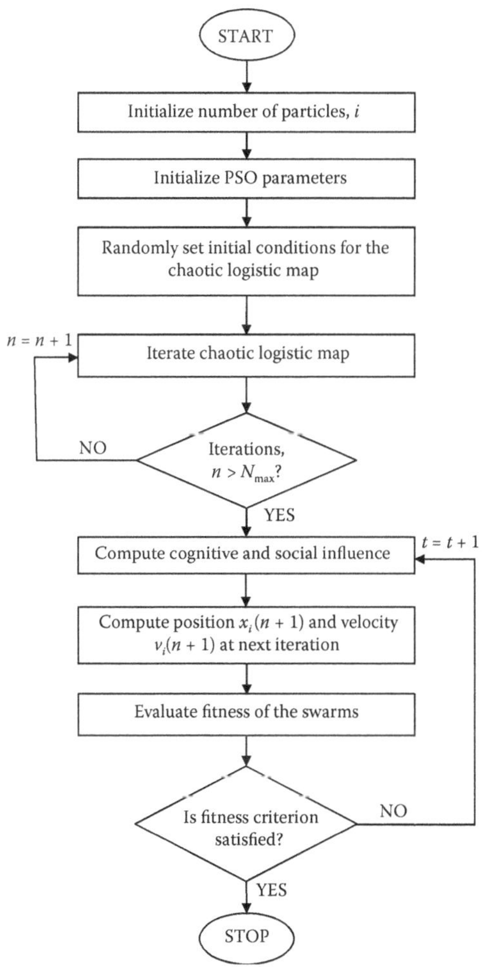

FIGURE 9.3 Flowchart for the Ch-PSO algorithm.

9.6 CHAOTIC DE

Specific modifications were performed in the DE algorithm to enhance its diversification capabilities by the addition of the chaotic component. First, the population of vectors, x_i^{G}, was generated. The consequent steps are similar to the regular DE algorithm where one principal parent x_i^{p} and three auxiliary parents x_i^{a} are randomly selected. Differential mutation is then performed and the mutated vector, V_i, is generated. The V_i is then recombined with x_i^{p} to generate child trial vector, x_i^{child}. The obtained x_i^{child} is used as the input to the chaotic logistic map (Jakobson, 1981). This chaotic logistic mapping is presented as follows:

$$N_i(t) = x_i^{\mathrm{child}}(t) \tag{9.30}$$

$$R_i(t) = \lambda N_i(t) \tag{9.31}$$

$$N_i(t+1) = R_i(t) N_i(t)\left[1 - N_i(t)\right] \tag{9.32}$$

$$R_i(t+1) = R_i(t) + \lambda' \tag{9.33}$$

where $N(t)$ and $R(t)$ are variables in the logistic chaotic map, and λ' and λ are relaxation constants specified by the user. Then the logistic mapping is continued until a specific number of iterations is satisfied.

The final value at maximum number of iterations of $N(t_{\mathrm{max}})$ is incorporated into the child trial vector, x_i^{child}. Hence, the child trial vector, x_i^{child}, undergoes another round of mutation by the chaotic map. Next, the "knock-out" competition for next-generation survival selection is performed. The fitness function for the child trial vector, x_i^{child}, is evaluated. Thus, another variant of the DE algorithm with enhanced diversification capabilities called the CDE is presented. The flowchart and the algorithm are given in Figure 9.4 and Algorithm 9.4 respectively:

Algorithm 9.4: Chaotic Differential Evolution (CDE)

Step 1: Initialize population size
Step 2: Randomly initialize the population vectors, x_i^{G}
Step 3: Randomly select one principal parent, x_i^{p}
Step 4: Randomly select three auxiliary parents, x_i^{a}
Step 5: Perform differential mutation and generate mutated vector, V_i
Step 6: Recombine V_i with x_i^{p} to generate child trial vector, x_i^{child}
Step 7: Iterate chaotic logistic map
Step 8: If $n > N_{\mathrm{max}}$, proceed to next step else go to step 7
Step 9: Evaluate fitness of the new x_i^{child}
Step 10: If the fitness criterion is satisfied halt and print solutions, else proceed to step 3

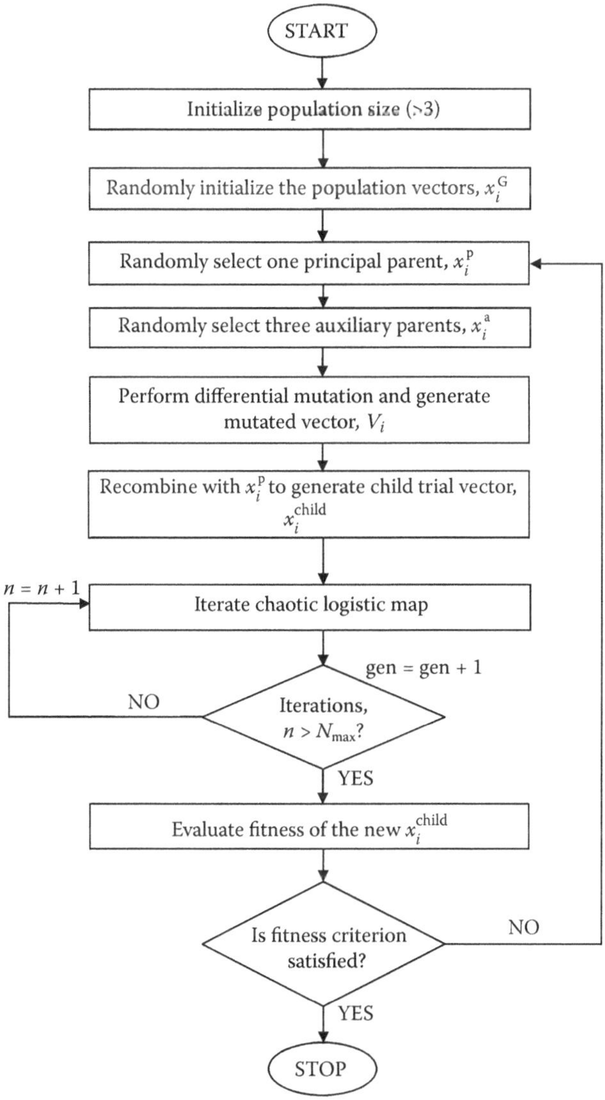

FIGURE 9.4 Flowchart for the CDE algorithm.

9.7 GAME-THEORETIC DE

In past years, game theory has been used extensively to model and explain systems with conflicting scenarios (e.g., biology, economics, and other social sciences) (Axelrod, 1984; Myerson, 1991; Weibull, 1995). The single primary proposition of game theory is:

> The rational pursuit of individual self-interest collectively produces a self-defeating outcome.

Using this as the core idea, a more general form of game theory, the evolutionary game theory (EGT), was produced. In EGT, the system being analyzed consists of a large number of players that randomly interact in pairs. The EGT has proven to be extremely efficient in modeling emergence and dynamics involving collective behaviors. Some elements of the EGT are used to modify and improve the conventional DE technique. The EGT scheme functions as a regulating subsystem in the DE approach resulting in the game-theoretic differential evolution (GTDE) approach. In the proposed technique, the offspring and the parents are engaged competitively to achieve the user-specified aim in each generation. The aim is to obtain highly optimal solutions that satisfy all the required constraints. The EGT-based strategies are used to construct regulatory subsystems for the DE algorithm. The EGT strategies are stated as follows:

- Cooperative move—Reducing the mutation factor of the child vector
- Defective move—Modifying relation to increase degree of mutation for the child vector.

The algorithm for the GTDE approach employing the mentioned strategies are shown in Algorithm 9.5 and Figure 9.5.

Algorithm 9.5: Game-Theoretic Differential Evolution (GTDE)

Step 1: Initialize individual size and initialize the population vectors, x_i^{G}
Step 2: Select one principal parent, x_i^{p}, and three auxiliary parents, x_i^{a}, at random
Step 3: Evaluate fitness of current solution relative to the previous solution
Step 4: IF current fitness, FIT(gen), is better than previous fitness, FIT(gen−1)
Employ cooperative strategy between parent and child vector
ELSE
Employ defective strategy between parent and child vector
Step 5: Perform differential mutation and generate mutated vector, V_i
Step 6: Recombine V_i with x_i^{p} to generate child trial vector, x_i^{child}
Step 7: Perform "knock-out" competition for next-generation survival selection
Step 8: If the fitness criterion is satisfied and gen = $T_{\max}$, halt and print solutions, else proceed to step 3

9.8 OPTIMIZATION RESULTS

The solution sets which are the approximations of the Pareto frontier were obtained using the DE, PSO, and GSA methods. The approximate Pareto frontiers obtained using the DE, PSO, and GSA algorithms are shown in Figures 9.6 through 9.8. The level of dominance gauged by the HVI for the entire Pareto frontier produced by the DE, PSO, and GSA techniques is given in Figure 9.9.

The frontier produced by the GSA algorithm in Figure 9.8 seems to be conglomerated and localized at certain portions of the objective space. This spacing property is seen to heavily influence the ability of the algorithm to approximate the Pareto

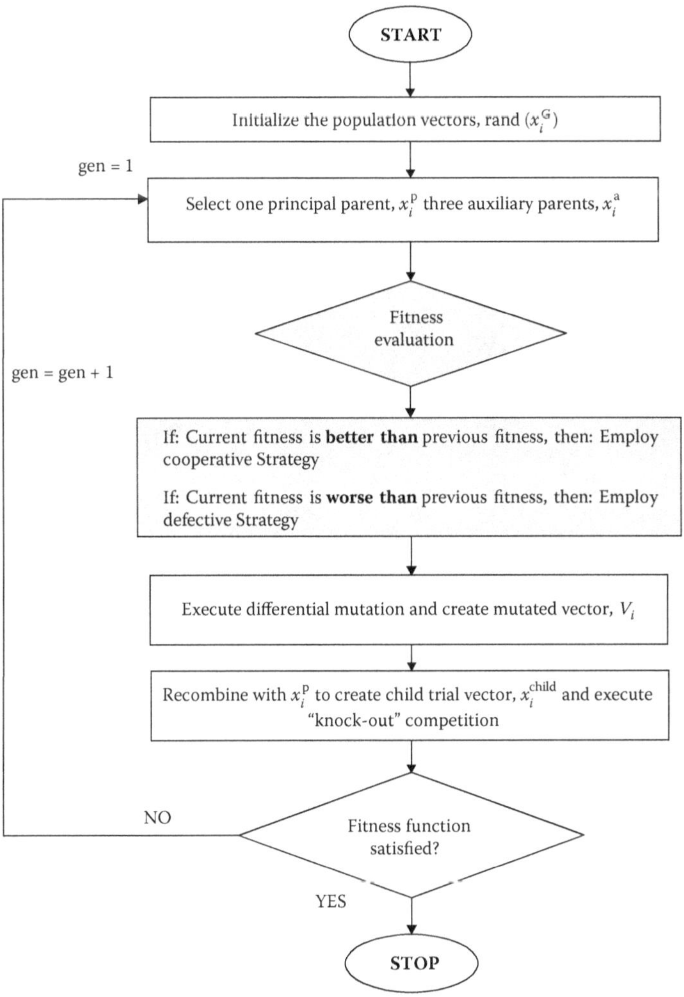

FIGURE 9.5 Flowchart for the GTDE approach.

frontier. Localized solutions of the frontier such as the ones produced by the GSA algorithm miss out on certain solutions in the objective space. Therefore, this causes the GSA algorithm to have a lower HVI as compared to the DE and PSO algorithms. Hence, as seen in Figure 9.9, this causes the approximated Pareto frontier produced by the GSA algorithm to become less dominant as compared to the one produced by the DE and PSO algorithms. It can be observed that the PSO algorithm produces the most dominant frontier followed by the DE and GSA algorithms sequentially. The PSO algorithm dominates the DE algorithm by 18.08%. To obtain a better understanding on the two most efficient algorithms (DE and PSO), further analysis on the solution diversity characteristics is conducted using the sigma diversity metrics. The diversity value of the frontier generated by the PSO, GSA, and DE techniques are 0.762, 0.75, and 0.6071, respectively. In the diversity analysis,

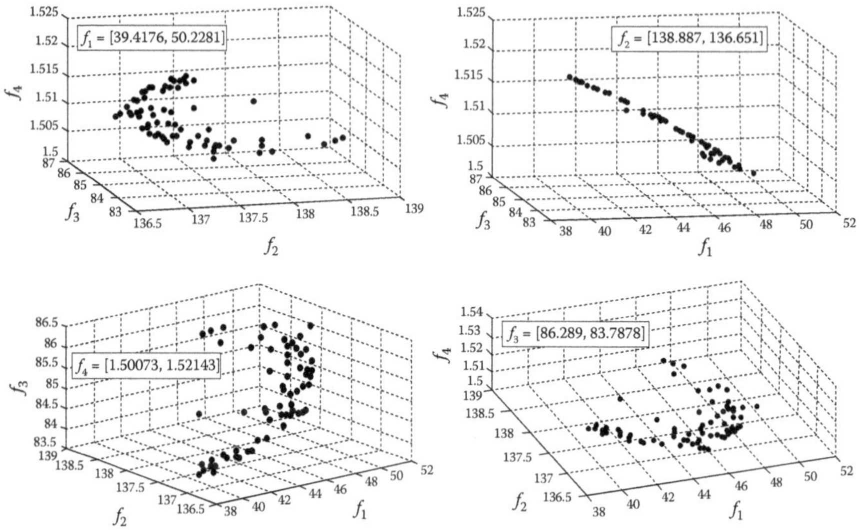

FIGURE 9.6 Pareto frontiers of the objectives obtained by the DE method.

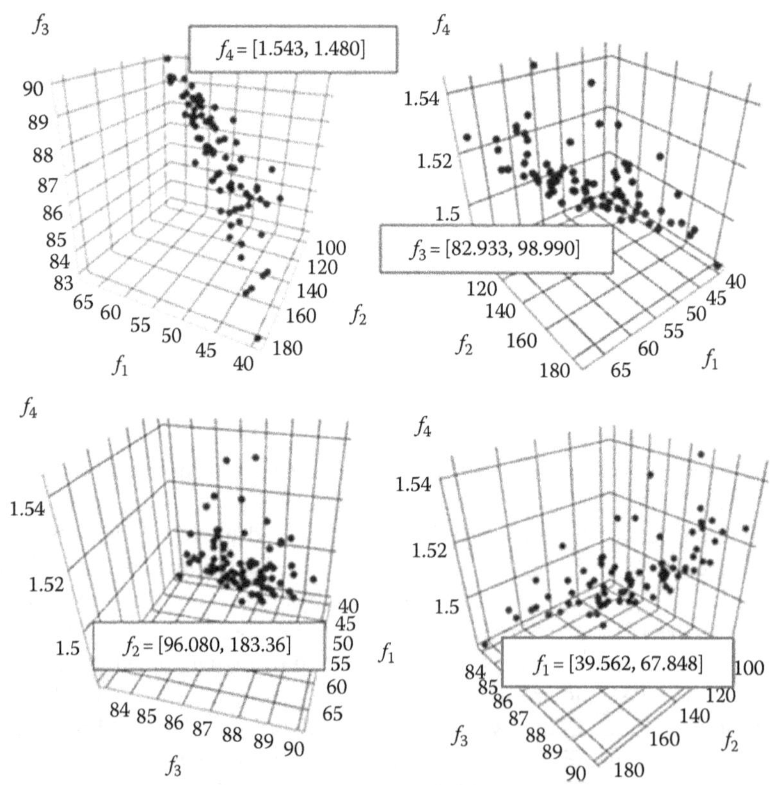

FIGURE 9.7 Pareto frontiers of the objectives obtained by the PSO method.

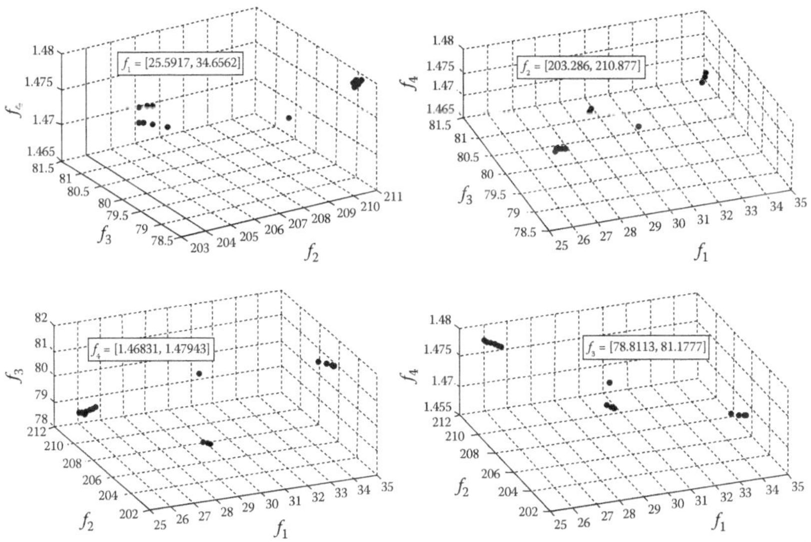

FIGURE 9.8 Pareto frontiers of the objectives obtained using GSA.

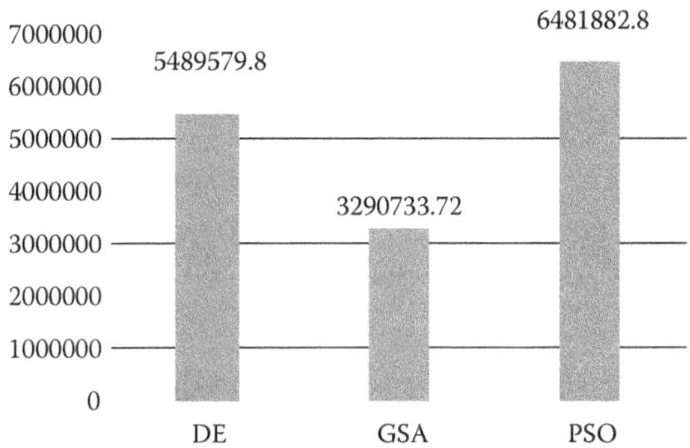

FIGURE 9.9 Dominance of Pareto frontiers generated by the DE, GSA, and PSO techniques.

the GSA technique would have to be excluded since, as seen in Figure 9.8, the solutions produced by GSA technique are localized although they are diversely spread. This gives a false impression that the solutions have a high diversity covering many regions of the objective space. This issue could be identified as one of the drawbacks of the sigma diversity metric whereby it cannot flag off localized solutions as compared to diversely spread solutions. Thus, looking at the dominance levels of the PSO and DE techniques in Figure 9.9, it can be clearly seen that the degree of frontier diversity corresponds to the level of frontier dominance. Therefore, to further boost the performances of these two effective techniques, the enhanced algorithms, CPSO, CDE, and GTDE, are implemented to the green sand molding problem. The frontiers generated by these techniques are given in Figures 9.10 through 9.12.

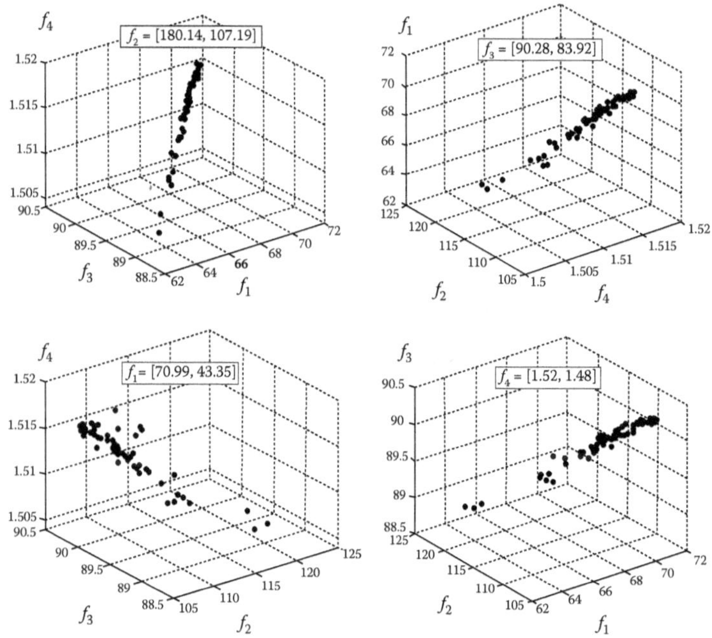

FIGURE 9.10 Pareto frontiers of the objectives obtained using CPSO.

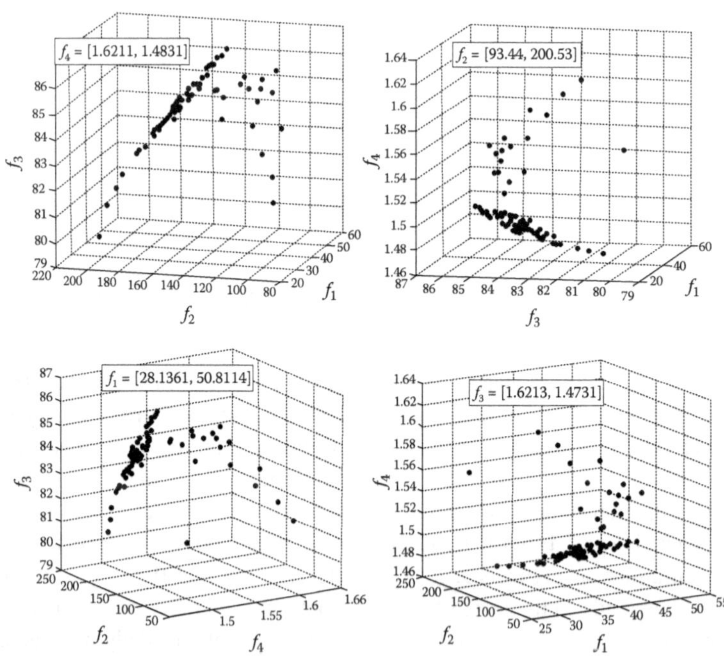

FIGURE 9.11 Pareto frontiers of the objectives obtained using CDE.

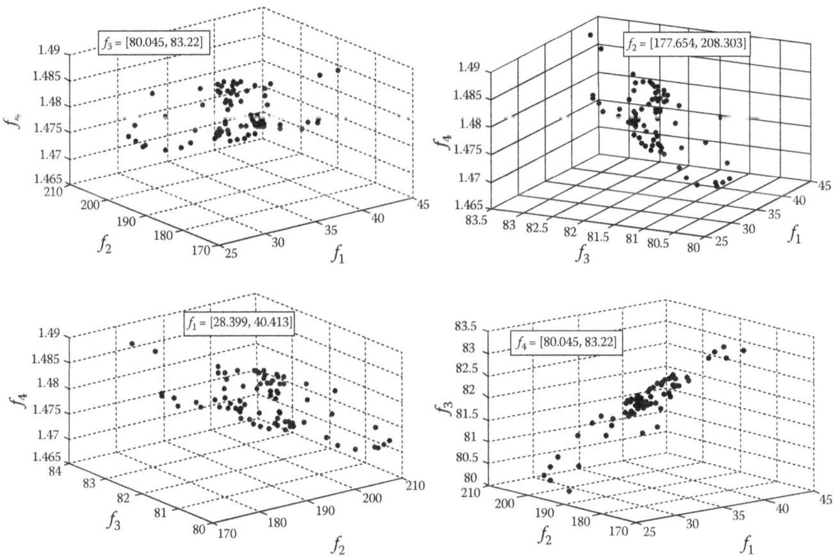

FIGURE 9.12 Pareto frontiers of the objectives obtained using GTDE.

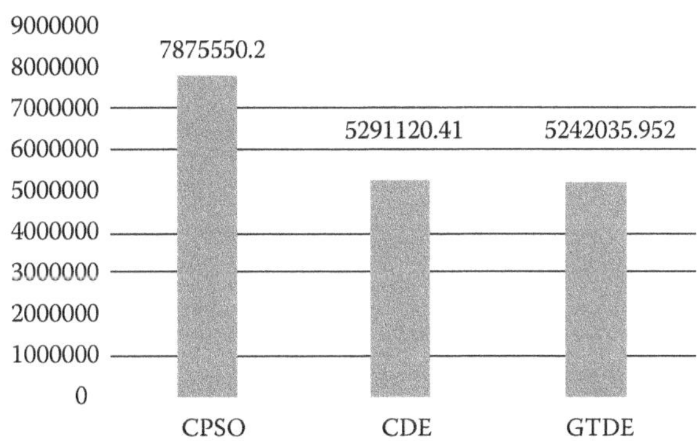

FIGURE 9.13 Dominance of Pareto frontiers generated by CPSO, CDE, and GTDE.

The level of dominance gauged by the HVI for the entire Pareto frontiers produced by the CPSO, CDE, and GTDE techniques are given in Figure 9.13.

Based on Figure 9.12, it can be stated that the CPSO technique produces the most dominant Pareto frontier since it optimizes all the objectives to an equal extent as compared to the other enhanced algorithms. The frontier diversities of the CPSO, CDE, and GTDE techniques are 0.778, 0.75, and 0.75, respectively. Therefore, similar to the non-enhanced algorithms, the diversity of the solution spread influences the levels of dominance of the Pareto frontier. In this case, the CDE and the GTDE techniques seem to produce an equally diverse frontier. The CPSO has the most diversely spread as well as dominant Pareto frontier as compared to all the other approaches employed to solve the green sand molding problem.

9.9 PERSPECTIVES ON OPTIMIZATION RESULTS

The MO optimization of the green sand mold system was performed via an exhaustive and rigorous implementation of a series of metaheuristic approaches. These explorations and numerical experimentations have proven that algorithmic enhancements are necessary to boost the performance of metaheuristics during optimization. In addition, by the employment of measurement metrics, critical features of the solution space could be ascertained. Besides, knowledge on the interaction between the individual algorithms and the problem at hand may also be extracted using such measurement approaches. The information gained regarding these features may prove to be helpful to the decision maker when considering algorithmic enhancement options or other improvements related to the optimization procedures.

REFERENCES

Axelrod, R. (1984). *The evolution of cooperation.* New York: Basic Books, Inc.

Beume, N., Naujoks, B., & Emmerich, M. (2007). SMS-EMOA: Multiobjective selection based on dominated hypervolume. *European Journal of Operational Research, 181*(3), 1653–1669.

Colorni, A., Dorigo, M., & Maniezzo, V. (1991). Distributed optimization by ant colonies. *Proceedings of the First European Conference of Artificial Intelligence* (pp. 134–142). Paris, France: Elsevier Publishing.

Das, I., & Dennis, J. E. (1998). Normal-boundary intersection: A new method for generating the Pareto surface in nonlinear multicriteria optimization problems, *SIAM Journal of Optimization, 8*(3), 631–657.

Deb, K., Pratap, A., Agarwal, S., & Meyarivan, T. (2002). A fast and elitist multiobjective genetic algorithm: NSGA-II. *IEEE Transactions on Evolutionary Computation, 6*(2), 182–197.

Eschenauer, H., Koski, J., & Osyczka, A. (1990). *Multicriteria design optimization.* Berlin, Germany: Springer-Verlag.

Farhang-Mehr, A., & Azarm. S. (2002). Diversity assessment of Pareto optimal sets: An entropy approach. In *IEEE Proceedings World Congress on Computational Intelligence (CEC'02)*, May 12–17, 2002, Honolulu, Hawaii.

Flake, G. W. (1998). *The computational beauty of nature: Computer explorations of fractals, chaos, complex systems, and adaptation* (pp. 469–482). Cambridge, MA: MIT Press.

Fleischer, M. (2003). The measure of Pareto optima. Applications to multi-objective metaheuristics. In *Conference on Evolutionary Multi-Criterion Optimization (EMO 2003)*, April 8–11, 2003 (pp. 519–533), Faro, Portugal. Springer.

Grosan, C. (2003). Performance metrics for multiobjective optimization evolutionary algorithms. In *Proceedings of Conference on Applied and Industrial Mathematics* (*CAIM*), Oradea.

Igel, C., Hansen, N., & Roth, S. (2007). Covariance matrix adaptation for multi-objective optimization. *Evolutionary Computation, 15*(1), 1–28.

Jakobson, M. (1981). Absolutely continuous invariant measures for one-parameter families of one-dimensional maps. *Communications on Mathematical Physics, 81*, 39–88.

Knowles, J., & Corne, D. (2003). Properties of an adaptive archiving algorithm for storing nondominated vectors. *IEEE Transactions on Evolutionary Computation, 7*(2), 100–116.

Lorenz, E. N. (1963). Deterministic non-periodic flow. *Journal of the Atmospheric Sciences, 20*(2), 130–141.

Luyben, M. L., & Floudas, C. A. (1994). Analyzing the interaction of design and control. 1. A multiobjective framework and application to binary distillation synthesis. *Computers and Chemical Engineering*, *18*(10), 933–969.

Mostaghim, S., & Teich, J. (2003). Strategies for finding good local guides in multiobjective particle swarm optimization. In *IEEE Swarm Intelligence Symposium* (pp. 26–33), Indianapolis, IN.

Myerson, R. B. (1991). *Game theory: Analysis of conflict*. Cambridge, MA: Harvard University Press.

Okabe, T., Jin, Y., & Sendho, B. (2003). A critical survey of performance indices for multiobjective optimization. In *IEEE Proceedings, World Congress on Computational Intelligence (CEC'03)* (pp. 878–885), Canberra, Australia.

Parappagoudar, M. B., Pratihar, D. K., & Datta, G. L. (2007). Non-linear modeling using central composite design to predict green sand mold properties. *Proceedings IMechE B Journal of Engineering Manufacture*, *221*, 881–894.

Rashedi, E., Nezamabadi-pour, H., & Saryazdi, S. (2009). GSA: A gravitational search algorithm. *Information Sciences*, *179*, 2232–2248.

Statnikov, R. B., & Matusov, J. B. (1995). *Multicriteria optimization and engineering*. New York: Chapman and Hall.

Surekha, B., Kaushik, L. K., Panduy, A. K., Vundavilli, P. R., & Parappagoudar, M. B. (2012). Multi-objective optimization of green sand mold system using evolutionary algorithms. *The International Journal of Advanced Manufacturing Technology*, *58*(1–4), 9–17.

Sushil, K., Satsangi, P. S., & Prajapati, D. R. (2010). Optimization of green sand casting process parameters of a foundry by using Taguchi method. *International Journal of Advanced Manufacturing Technology*, *55*, 23–34.

Triantaphyllou, E. (2000). *Multi-criteria decision making: A comparative study* (p. 320). Dordrecht, The Netherlands: Kluwer Academic Publishers (now Springer).

Weibull, J. W. (1995). *Evolutionary game theory*. Boston: MIT Press.

Zitzler, E. (1999). *Evolutionary algorithms for multiobjective optimization: Methods and applications*. Aachen, Germany: Shaker Verlag, TIK-Schriftenreihe Nr. 30, Diss ETH No. 13398. Zurich: Swiss Federal Institute of Technology (ETH).

Zitzler, E., & Thiele, L. (1998). Multiobjective optimization using evolutionary algorithms—A comparative case study. In A. E. Eiben, T. Bäck, M. Schoenauer, & H. P. Schwefel (Eds.), *Parallel problem solving from nature, V* (pp. 292–301). Berlin, Germany: Springer.

Zitzler, E., Thiele, L., Laumanns, M., Fonseca, C. M., & Grunert da Fonseca, V. (2003). Performance assessment of multiobjective optimizers: An analysis and review. *IEEE Transactions on Evolutionary Computation*, *7*(2), 117–132.

10 Multiobjective Optimization of Resin-Bonded Sand Core Properties

In the foundry industry, binders are introduced into the mold to improve casting accuracy and the physical properties (e.g., mold hardness and strength) of the mold. Currently, organic binders are gradually replacing the hot-box process which uses a thermoset (also known as catalytic binders). Resin-bonded sand mold systems are thus categorized as a cold-box molding process which uses organic binders.

10.1 RESIN-BONDED SAND CORE SYSTEMS

In mold development systems, the mold quality (which has multiple criteria) is directly influenced by the molding conditions such as percentage of constituents, curing time, and number of strokes. The effective optimization of this process would require a comprehensive understanding of multiobjective (MO) optimization, cold-box molding, and metaheuristics. Over the past years, many research works have been directed toward resin-bonded sand mold systems due to the following factors:

- Excellent adaptive capabilities in a dynamics environment
- Good compliance to environmental factors
- High casting quality (Ireland, Chang, & Kroker, 2002).

This approach is power saving and thus highly suitable for large-scale production (Weihua, Yingmin, Xueliang, & Xiuling, 2008). In most cases, resin-bonded sand mold systems show superb flow behavior. Nevertheless, some introduction of vibration and compaction during the molding process is required. Casting characteristics are very much influenced by the mold properties which are in effect dependent on the process parameters during the mold development process (Parappagoudar, Pratihar, & Datta, 2007). In this work, the MO optimization model employed for the resin-bonded sand system was developed by Surekha, Hanumantha Rao, Mohan Rao, Vundavilli, and Parappagoudar (2012). This model was based on a sand mold where phenol formaldehyde was used as a binder and tetrahydrophthalic anhydride was used as a hardener.

The model describing the responses and the outputs of the optimization process was developed by Surekha et al. (2012). In that work, the mechanical properties of the quartz-based resin-bonded sand core system was modeled using Mamdani-based fuzzy logic (Mamdani & Assilian, 1975) and genetic algorithm (GA) approaches. The MO representation of the optimization model which consists of four objectives as developed in Surekha et al. (2012) is as follows:

$$\begin{aligned}&\text{Maximize} \rightarrow \text{Permeability}, f_1\\&\text{Maximize} \rightarrow \text{Compression Strength}, f_2\\&\text{Maximize} \rightarrow \text{Tensile Strength}, f_3\\&\text{Maximize} \rightarrow \text{Shear Strength}, f_4\\&\text{subject to } \textit{process} \text{ constraints.}\end{aligned} \tag{10.1}$$

The response parameters are: *A*, percentage of resin (%); *B*, percentage of hardener (%); *C*, number of strokes; and *D*, curing time (minutes). The final formulation of nonlinear regression model developed by Surekha et al. (2012) and the associated Constraints are given as follows:

$$\begin{aligned}f_1 = &-333.77 + 614.73A - 27.435B + 630.36C\\&-18.97D - 168.98A^2 + 0.239B^2 - 76.08C^2\\&+0.111D^2 + 2.827AB + 0.575AC + 0.047AD\\&-0.7701BC + 0.1323BD - 0.1883CD\end{aligned} \tag{10.2}$$

$$\begin{aligned}f_2 = &\,2765.36 + 877.869A - 112.778B - 731.934C + 17.9222D\\&-357.829A^2 + 0.983456B^2 + 52.2310C^2 - 0.0276946D^2\\&+14.6571AB + 96.8495AC - 3.74068AD + 7.62554BC\\&-0.096084BD - 1.27093CD\end{aligned} \tag{10.3}$$

$$\begin{aligned}f_3 = &-354.406 + 211.418A + 17.3611B + 96.7916C + 2.78503D - 44.7516A^2\\&-0.173996B^2 - 10.6696C^2 - 0.026223D^2 - 2.08868AB + 6.05542AC\\&+0.197646AD + 2.07847BC - 0.078904BD + 1.18561CD\end{aligned} \tag{10.4}$$

$$\begin{aligned}f_4 = &\,318.163 + 726.696A + 33.3432B - 721.381C + 2.40622D - 210.057A^2\\&-0.189623B^2 + 80.1788C^2 + 0.000987D^2 - 1.89739AB + 49.8702AC\\&-0.32471AD - 1.70998BC - 0.07323BD + 0.306223CD\end{aligned} \tag{10.5}$$

$$A \in [1.5, 2.5],\ B \in [30, 50],\ C \in [3, 5],\ D \in [60, 100] \tag{10.6}$$

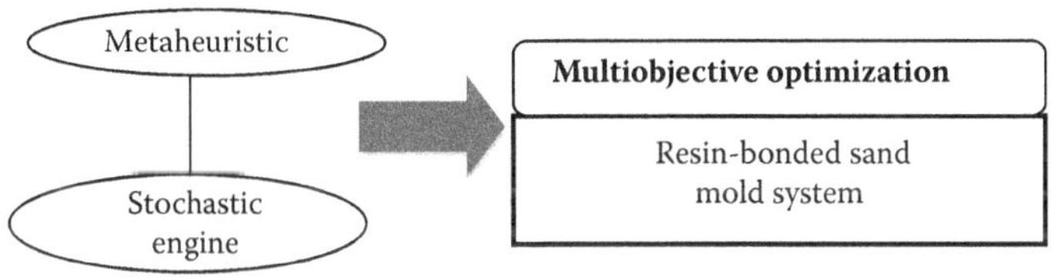

FIGURE 10.1 Graphical representation of the main concepts.

In this chapter, analyses involving the effects of non-Gaussian stochastic engines on the performance of metaheuristics are presented. Details involving the implementation of two metaheuristics (bacteria foraging algorithm [BFA] and the particle swarm optimization [PSO]) on a real-world engineering problem is explored. The influence of the three types of stochastic engines (chaos based, Gaussian type, and non-Gaussian type) on the resin-bonded sand mold system is studied. The behavior and the results generated by these techniques are investigated by employing the hypervolume indicator (HVI) and the average explorative rate (AER) metric. An illustration of the main concepts presented in this chapter is given in Figure 10.1.

10.2 SWARM INTELLIGENCE

In agent-based algorithms (Wu & Chen, 2009; Wernick, Yang, Brankov, Yourganov, & Strother, 2010) a swarm is a group of artificial organisms or agents that behave interactively to achieve some defined goal. This form of interaction gives the individuals higher capabilities as well as better efficiency in achieving some arbitrary goal as compared to a single individual. Many computational studies of swarm intelligence (SI) have been conducted over the past years involving the flocking of birds, the foraging of ants, motions of schools of fish, as well swarming of bees and wasps (Cui & Gao, 2012).

Individuals in the swarm subsequently update and improve their information about the environment as they move toward their goal. Each individual is thus equipped with individual cognition (self-analysis) as well as social cognition (social analysis). As the swarm moves, the individuals record their fitness values (with respect to some objective function). By individual and social cognition the swarm gravitates toward the fittest regions in the objective space by improving its fitness values from the previous movements as the motion progresses. Hence, SI-based algorithms are a class of metaheuristic algorithms.

Thus, among all the state-of-the-art approaches for solving highly complex optimization problems, SI stands as one of the most favored strategies. This is mainly due to its effectiveness in the search process and its efficiency in terms of computational time (Liu, Passino, & Simaan, 2002). Some of the most common SI-based approaches are PSO (Kennedy & Eberhart, 1995), cuckoo search (CS) (Yildiz, 2013), ant colony optimization (ACO) (El-Wahed, Mousa, & Elsisy, 2008), and BFA (Passino, 2002).

10.3 BACTERIA FORAGING ALGORITHM

In the past, one of the most popular techniques employed for solving constrained nonlinear optimization problems is PSO. However, in recent times, other strategies such as BFA have become attractive for this purpose. BFA's computational performance has been proven to be as good as and at times better than other SIs as well as evolutionary techniques (Lin & Liu, 2006; Niu, Zhu, He, & Zeng, 2006). Inspired by the natural behavior of the *Escherichia coli* bacterium searching for nutrients in the environment, Passino (2002) proposed the BFA for solving complex optimization problems. The central premise of this approach is that the bacteria tries to maximize its energy per unit time spent during foraging for nutrients while simultaneously evading noxious substances.

The dynamics of bacteria foraging is directly influenced by evolutionary biology. Thus, bacteria with successful foraging strategies would stand a better chance in propagating their genetic makeup as compared to bacteria with poor strategies. This way, bacteria at successive generations always contain improved foraging strategies relative to past generations and the strategies continually improve as they go along reproducing. Due to such progressive behavior, many research studies were targeted to model bacteria foraging dynamics as an optimization process. The central theme of foraging viewed from this perspective is that the organisms conduct the search in such a way that they maximize the energy they obtain from the nutrients in minimal time during foraging.

Foraging efforts vary according to the species of the organism and the environment where the foraging is taking place. For instance, herbivores would find it easier to locate food as compared to carnivores in any habitat. As for the environmental factor, the distribution of nutrients in desert or tundra conditions is sparser in contrast with the nutrient-rich tropical jungles. Design of effective and efficient search strategies for nutrient foraging which respects the previous constraints is critical for the long-term well-being of any organism. Another important factor to be considered for the design of effective search strategies is the type of nutrient. The type of nutrient will influence the fractionalization and planning of the strategy (O'Brien, Browman, & Evans, 1990). For instance, consider a case where the nutrient is stationary but hidden in a hard shell (e.g., eggs). Then the organism would have designed the foraging strategy in such a way that it (1) searches for the shell, (2) evades the parent, (3) breaks the shell, (4) consumes the nutrient, and (5) then escapes the nutrient location or nest before it gets attacked/killed.

In many organisms, synergetic foraging strategies are observed to emerge in nature (e.g., ants, bees, and termites). These organisms create communication mechanisms that enable them to share information about the foraging efforts led by each individual in the group. Such mechanisms provide the capability of the organisms to conduct "group/swarm foraging." Group foraging provides these organisms with a plethora of advantages such as increased protection against predators and enhanced hunting/foraging strategies. These advantageous traits increase the organism's chances for finding nutrients in good time. Besides synergetic strategies for foraging, other strategies such as cooperative building (Scott Turner, 2011), group defense (Schneider & McNally, 1992), and other cooperative group behaviors are common in nature.

10.4 MAIN MECHANISMS OF BFA

In the BFA, four main levels of loops are present in the technique (chemotaxis, swarming, reproduction, and elimination-dispersal loops). These loops manage the main functional capabilities of the BFA. Each of the mentioned loops is designed according to bacteria foraging strategies and principles from evolutionary biology. These loops are executed iteratively until the total number of iterations, N_T, is satisfied. Each of the main loops may be iterated until some fitness condition is satisfied or until a user-defined loop cycle limit (chemotaxis [N_c], swarming [N_s], reproduction [N_r], and elimination-dispersal [N_{ed}]) is reached.

In chemotaxis, the bacteria with the use of their flagella swim and tumble toward the nutrient source. The swimming mode allows bacterium motion in a fixed direction while the tumbling mode enables the bacterium to augment its search direction accordingly. Applied in tandem, these two modes give the bacterium capability to stochastically move toward a sufficient source of nutrient. Thus, computationally, chemotaxis is presented as follows:

$$\theta^i(j+1,k,l,m)=\theta^i(j,k,l,m)+C(i)\frac{\Delta(i)}{\sqrt{\Delta(i)\Delta^T(i)}} \tag{10.7}$$

where $\theta^i(j+1, k, l, m)$ is the ith bacterium at the jth chemotactic step, kth swarming step, lth reproductive step, and mth elimination-dispersal step. $C(i)$ is the size of the step taken in a random direction which is fixed by the tumble, and $\Delta \epsilon$ [−1,1] is the random vector.

In the swarming phase, the bacterium communicates with the entire swarm regarding the nutrient profile it mapped during its movement. The communication method adopted by the bacterium is cell-to-cell signaling. In *E. coli* bacteria, aspartate (attractant) is released by the cells if they are exposed to high amounts of succinate (substrate). This causes the bacteria to conglomerate into groups and hence move in a swarm of high bacterial density. The swarming phase is mathematically presented as follows:

$$J(\theta,P(j,k,l,m))=\sum_{i=1}^{S}\left[-D_{att}\exp\left(-W_{att}\sum_{m=1}^{P}\left(\theta_m-\theta_m^i\right)^2\right)\right] + \sum_{i=1}^{S}\left[-H_{rep}\exp\left(-W_{rep}\sum_{m=1}^{P}\left(\theta_m-\theta_m^i\right)^2\right)\right] \tag{10.8}$$

where $J(\theta, P(j, k, l, m))$ is the computed dynamic objective function value (not the real objective function in the problem); S is the total number of bacteria; P is the number of variables to be optimized (embedded in each bacterium); and H_{rep}, W_{rep}, H_{att}, and W_{att} are user-defined parameters.

During reproduction, the healthy bacteria or the bacteria which are successful in securing a high degree of nutrients are let to reproduce asexually by splitting into two. Bacteria which do not manage to perform according to the specified criteria are eliminated from the group and thus not allowed to reproduce causing their genetic propagation (in this case their foraging strategies) to come to a halt. Due to this cycle, the amount of individual bacteria in the swarm remains constant throughout the execution of the BFA.

Catastrophic events in an environment such as a sudden change in physical/chemical properties or rapid decrease in nutrient content could result in death of a population of bacteria. Such events can cause bacteria to be killed and some to be randomly dispersed to different locations in the objective space. These events which are set to occur in the elimination/dispersal phase help to maintain swarm diversity to make sure the search operation is efficient. The flowchart depicting the execution steps of the BFA technique is given in Figure 10.2 while the pseudo-code for the BFA is as follows:

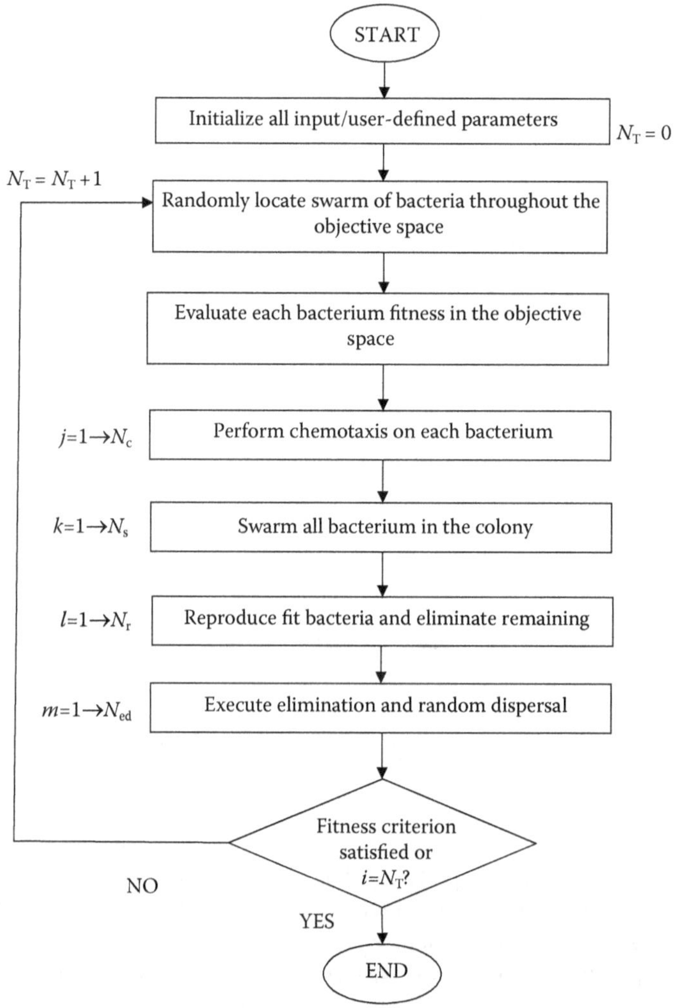

FIGURE 10.2 The BFA technique.

```
START PROGRAM
Initialize all input parameters (S, P, H_rep, W_rep, H_att, W_att, N_T,
N_c, N_r, N_s, N_ed)
Generate a randomly located swarm of bacteria throughout the
objective space
Evaluate bacteria fitness in the objective space
For i=1 → N_T do
        For l=1 → N_r do
                For m=1 → N_ed do
                        For j=1 → N_c do
                                For k=1 → N_s do
                                Perform chemotaxis-
                                bacterium swim and tumble
                                until maximum fitness/loop
                                cycle limit is reached
                                Perform swarming—bacterium
                                swarm until maximum
                                fitness/loop cycle limit
                                is reached
                                End For
                        End For
                If bacterium healthy/maximally fit then
                split and reproduce
                Else eliminate remaining bacterium
                End For
        Execute catastrophic elimination by assigning some
        probability of elimination to the swarm. Similarly
        disperse the remaining swarm randomly.
        End For
End For
END PROGRAM
```

10.5 BRIEF OVERVIEW OF BFA IN ENGINEERING

BFA has been seen to have many applications in engineering-related areas such as control systems, manufacturing technology, engineering design, power systems, and economic dispatch. For instance, in Mezura-Montes and Herńandez-Ocãna (2009), the authors proposed specific modifications to the BFA to increase its optimization capability. This modified version of the BFA was then applied to solve a constrained engineering design problem.

In Mezura-Montes, Portilla-Flores, and Herńandez-Ocãna (2014), the design optimization of a crank–rocker–slider (continuous variable transmission) system was carried out using the BFA. Since the design problem presents itself as a single- and double-objective problem, the original BFA was restructured by the authors to create a MO BFA. Both techniques were successfully implemented to optimize the mechanical design of the crank–rocker–slider system.

BFA has also been applied in the area of cellular manufacturing systems (Nouri & Hong, 2013) where the cell formation problem is solved while taking into account the cell load variations and the number of exceptional elements. In that work, the BFA was used to generate part families and machine cells. In Panda, Padhee, Sood, and Mahapatra (2009), the BFA was employed in rapid prototyping in manufacturing technology. In that work, the BFA was used for optimizing the process parameters employed for fused deposition modeling (FDM).

Besides manufacturing technologies, BFA has also been popular among experts in power systems. BFA has been utilized for optimizing economic load dispatch where the aim is to obtain the most optimal economic schedule for the power-generating units while simultaneously satisfying the load demands and load constraints (Vijay, 2011). In addition, BFA has been implemented for obtaining the optimal power flow (i.e., economic and efficient) in flexible alternating current transmission system (FACTS) devices (Ravi, Shilaja, Chitti Babu, & Kothari, 2014).

Currently, BFA has also been used for control system design. For instance in Bensenouci (2011), two PID controllers employed to control a power system were designed using the BFA technique. The gain settings for both controllers were obtained using BFA. The studies conducted in Bensenouci (2011) found that the designed controllers show good performance during power regulation, reference tracking, and parameter change tests. Another interesting application of BFA in control systems was carried out in the work of Rajinikanth and Latha (2012). An enhanced BFA technique was employed in that work to tune a PID controller. These controllers were employed to control chemical processes in real time involving a nonlinear spherical tank system. The enhanced BFA approach was then compared with the original BFA technique as well as other metaheuristic approaches such as PSO and GA. The investigations in Rajinikanth and Latha (2012) concluded that the controllers tuned using the enhanced BFA generated a smooth response for reference tracking and successfully maintained the level based on a reference signal during level control.

In the field of image processing, BFA was utilized in the study by Ben George and Karnan (2012) for brain image segmentation. In that work, Markov random field approach and the BFA technique were hybridized to perform image segmentation of brain magnetic resonance (MR) images. The segmentation was also carried out using a GA–Markov random field hybrid algorithm. These two strategies were then compared by the authors. It was found that the BFA–Markov random field approach generated very accurate results when compared against empirical data. It was also shown that the BFA-based hybrid provided a better characterization of the natural brain structure as compared to the GA-based hybrid technique. Besides segmentation, BFA has been applied to problems involving color image quantization. In Kumar (2013), the author implements the BFA for color image quantization and evaluates its performance on various file formats (bmp, png, and jpg). In that work, the author found that the BFA was more effective when applied to the jpg file format compared to other formats. The authors explained that since the jpg file format is designed for transmission and storage as compared to other file formats, the BFA quantizes files with this format more efficiently. The result after quantization with the BFA approach is that these files can be transmitted without compromising on their quality and can be stored over networks with less storage. A graphical overview of the BFA applications in engineering is given in Figure 10.3.

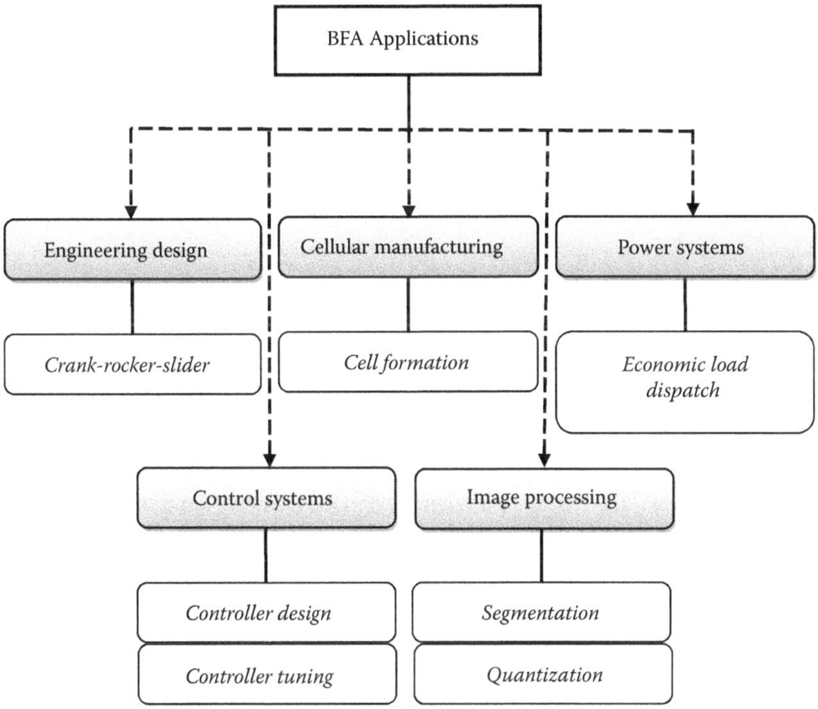

FIGURE 10.3 BFA applications in engineering systems.

10.6 STOCHASTIC ENGINES

Most optimization/search algorithms are incorporated with a stochastic engine that generates random values. Such components function to diversify the algorithm's search capability in the objective space. These stochastic engines also provide the algorithm with a "warm-start" for priming it for the search operation. Therefore, the type of probability distribution function (PDF) that generates the random values in the stochastic engine heavily influences the algorithm's search capability. In most metaheuristic algorithms, the stochastic engine produces random numbers according to a Gaussian PDF (Elamvazuthi, Ganesan, & Vasant, 2011). In other cases, researchers have employed other approaches to boost the stochastic engine by using chaos-based functions to couple with the Gaussian PDF (Afrabandpey, Ghaffari, Mirzaei, & Safayani, 2014). These approaches avoid the algorithm from getting stuck in the local optima. This sort of stagnation causes premature algorithmic convergence during computation. Thus, stochastic engines play a crucial role in the implementation of metaheuristics. Besides Gaussian stochastic engines, another class of distributions that has rarely been investigated with respect to metaheuristics are the non-Gaussian distributions. These are usually heavy-tailed profiles, for instance, Gumbel (Shirani, Härkegård, & Morinb, 2010), Weibull (Celik, 2004), Gamma (De Pascoa, Ortega, & Cordeiro, 2011), and Pareto (Roscoe & Diermanse, 2011) distributions. It has been known that many real-world applications (e.g., engineering, chemical, or economic systems) do not behave in a stochastically Gaussian manner but are prone to contain non-Gaussian fluctuations in their behavior. In such scenarios, a non-Gaussian approach for modeling or optimization may be a better solution method as compared to Gaussian-based strategies.

A random variable, $x \in X$, which is distributed with a mean, μ, and variance, σ^2, is said to be Gaussian or normally distributed when the PDF is as follows:

$$G_X(x) = \frac{1}{\sigma\sqrt{2\pi}} \exp\left[-\frac{1}{2}\left(\frac{x-\mu}{\sigma}\right)^2\right] \tag{10.9}$$

The Gaussian distribution is very general and widely applicable in various fields of studies for modeling real-valued random numbers (e.g., Brownian motion [Sottinen & Tudor, 2006] and Monte Carlo simulations [Corney & Drummond, 2004]). The standard normal distributions with $\mu = 0$ and $\sigma = 1$ are employed in the stochastic engines to generate random values in the metaheuristic. On the other hand, the Weibull distribution is a type of non-Gaussian distribution which is widely implemented in extreme value statistics. A two-parameter Weibull distribution function for a random variable $x \in X$ is defined as follows:

$$W(x) = 1 - \exp\left[-\left(\frac{x}{\lambda}\right)^k\right] \quad \text{for} \quad x \in R(0, +\infty) \tag{10.10}$$

where $W(x)$ is the Weibull distribution, k is the shape parameter, and λ is the scale parameter. It is required that the scale and shape parameter are positive for the Weibull distribution ($k > 0$ and $\lambda > 0$). For $\lambda = 1$, the Weibull distribution takes the form of the exponential distribution. It should be noted that the Weibull distributions around λ get smaller as the value of k increases. In this work, shape and scale parameters are set such that $\lambda = 1$ and $k = 5$. The Weibull distribution has been widely used in areas such as microbiology, information systems, and meteorology. Similar to the Weibull distribution, the Gamma distribution is another type of non-Gaussian distribution. The Gamma distribution is influenced by its shape, α, and rate, β, parameters. The Gamma distribution, $\gamma(x)$, for a random variable $x \in X$ is given as follows:

$$\gamma(x) = 1 - \sum_{i=0}^{\alpha-1} \frac{(\beta x)^i}{i!} \exp(-\beta x) \quad \text{for} \quad \forall i \in Z \text{ such that } \alpha > 0 \text{ and } \beta > 0 \tag{10.11}$$

The PDF of the Gamma distribution becomes near symmetrical if there is an increment in the shape factor and the mean as the skewness decreases. As the standard deviation of the distribution increases, the PDF gradually skews to the left and becomes heavy tailed. The Gamma distribution has been used successfully in climatology, insurance claim models, and risk analysis. Chaotic concepts could be employed to design a generator. A one-dimensional chaotic map could be used to initialize population of solutions by embedding the map into the random number generation component in the algorithm. The one-dimensional chaotic map, ψ_n, is represented as follows:

$$\psi_{n+1} = f(\psi_n) \tag{10.12}$$

The most widely studied one-dimensional map is the logistic map (Jakobson, 1981) which is as follows:

$$f(\psi_n) = r_n \psi_n (1 - \psi_n) \tag{10.13}$$

$$r_{n+1} = r_n + 0.01 \tag{10.14}$$

where $\psi_n \in [0,1]$ and $r_n \in [0,5]$. In this mapping, like all chaotic maps, the dynamics of the system varies for different sets of initial conditions (ψ_0 and r_0).

10.7 EXPLORATIVE RATE

In addition to the HVI, a novel metric, the AER is introduced for the purpose of measuring the thoroughness of the search operation carried out by the computational technique in the regions of the objective space. The AER performs online measurements successively during the execution of the computational technique. This metric measures the average amount of search region covered by the computational technique at each iteration. The AER can be computed by first determining the deviation of the objective function values at each iteration:

$$\delta_n = \left[\frac{f^{n+1}(x_i) - f^n(x_i)}{f^n(x_i)} \right] \tag{10.15}$$

where $f^n(x_i)$ is the objective function value at the nth iteration with x_i being the ith decision variables. Then the Heaviside Step Function is employed to return a value if the deviation, δ, is more than some predefined value, L.

$$H(\delta_n) = \begin{cases} 0 & \text{if } \delta_n < L \\ 1 & \text{if } \delta_n \geq L \end{cases} \tag{10.16}$$

where $H(\delta_n)$ is the Heaviside step function. The AER (E_R) is then computed as follows:

$$E_R = \sum_{n=1}^{N} \left(\frac{H(\delta_n)}{N} \right) \tag{10.17}$$

where n is the iteration count and N is the maximum number of iterations. Therefore, the larger the AER value, the more objective space is covered by the computational technique per iteration. This in effect results in a better search operation. It should be noted that when comparing computational techniques, the threshold value, L, must be consistent throughout the computational experiments.

10.8 COMPUTATIONAL EXPERIMENTS

10.8.1 BFA-Based Techniques

The BFA technique is executed with four of the stochastic engines discussed previously. The BFA technique equipped with the Gaussian, Weibull, and Gamma distributions are termed G-BFA, W-BFA, and γ-BFA respectively while the BFA coupled with the chaotic generator is called the Ch-BFA. The solution sets generated by the BFA variants are employed to construct the approximate Pareto frontier. For the approximation of the Pareto frontier, 53 solutions for various weights were obtained for each of the BFA variants employed in this work. Since the approaches are stochastic in nature, each solution point is selected by taking the best solution obtained from 10 independent runs of the algorithms (for each of the individual weights). The approximate Pareto frontier obtained using the BFA-based approaches are shown in Figures 10.4 through 10.7.

The degree of dominance which is obtained using the HVI for the Pareto frontiers produced by all four BFA variants is presented in Figure 10.8.

In Figure 10.8, it can be observed that the W-BFA generates the most dominant Pareto frontier followed by the G-BFA, Ch-BFA, and γ-BFA respectively. The frontier obtained using the W-BFA approach is more dominant than G-BFA, Ch-BFA, and γ-BFA by 6.156%, 38.385%, and 71.394%, respectively. It can be observed in Figure 10.5 that the frontier produced by the W-BFA is localized at the most optimal regions in the objective space. The G-BFA and the Ch-BFA generate solution set which is highly localized in certain regions (Figures 10.4 and 10.7). Thus, these approaches have very limited solution coverage which thus affects the overall dominance of the Pareto frontier. The γ-BFA on the other hand has a wide spread

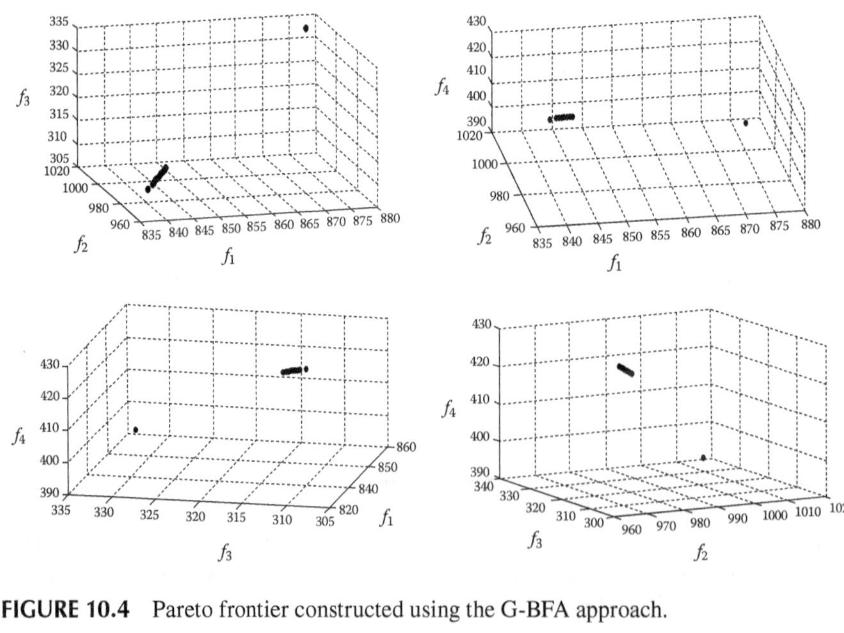

FIGURE 10.4 Pareto frontier constructed using the G-BFA approach.

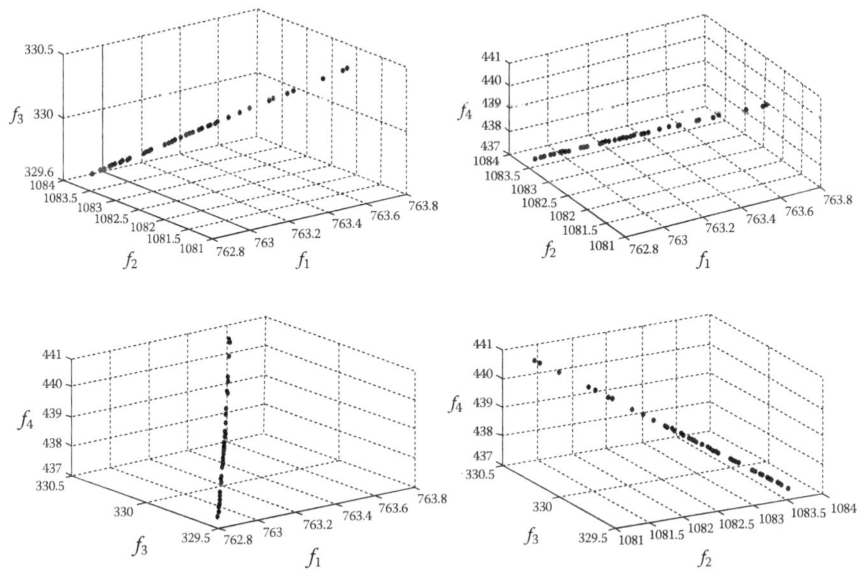

FIGURE 10.5 Pareto frontier constructed using the W-BFA approach.

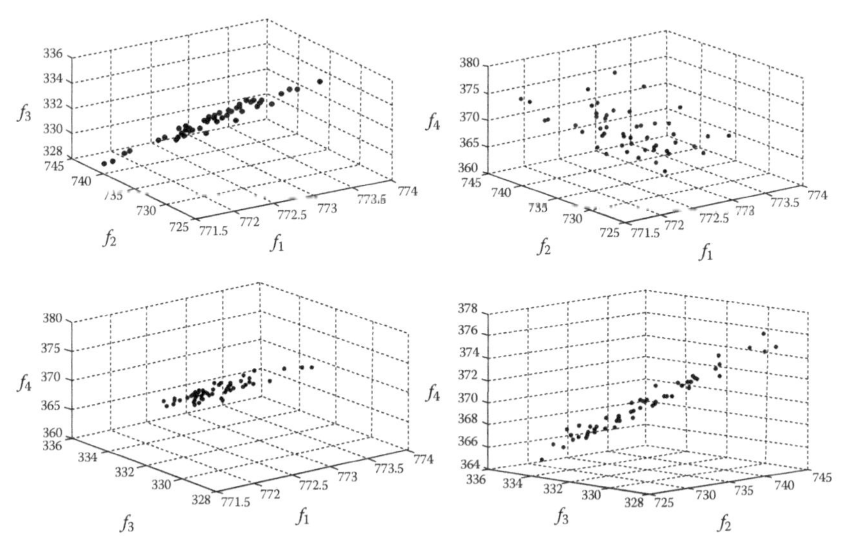

FIGURE 10.6 Pareto frontier constructed using the γ-BFA approach.

of solutions and hence high area of coverage (Figure 10.6). However, the solutions produced by the γ-BFA are not located on the optimal or dominant regions of the objective space. It can be seen in this work that wide coverage of solutions on the Pareto frontier is a critical criterion for Pareto dominance. Nevertheless, if the solutions are not located in the dominant/optimal regions in the objective space, the

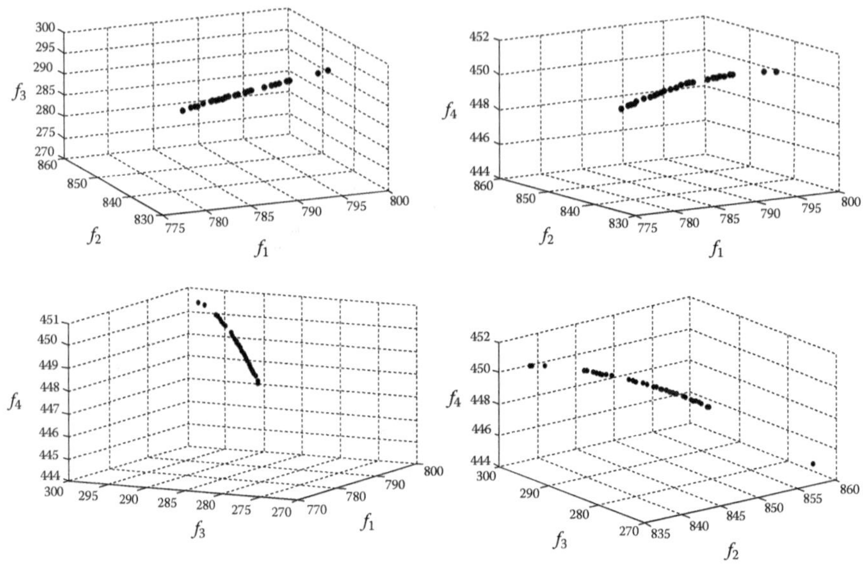

FIGURE 10.7 Pareto frontier constructed using the Ch-BFA approach.

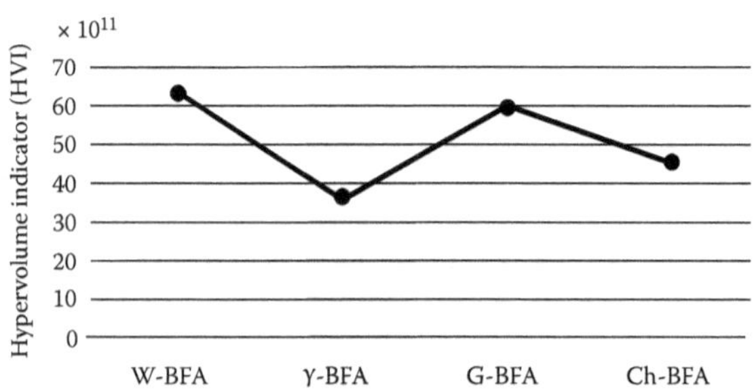

FIGURE 10.8 HVI value for the entire Pareto frontiers produced by the BFA variants.

overall frontier may not be highly dominant albeit widely spread. Throughout these executions, the Ch-BFA does not perform well as compared to the γ-BFA and the W-BFA (Figure 10.8). It is clear that for the sand mold system problem the chaotic component in the stochastic engine does not improve the performance of the metaheuristic approach. Similar to the degree of dominance of the entire Pareto frontier, the W-BFA outranks all the other BFA variants in terms of best individual solution.

With respect to best individual solution, the W-BFA outranks the γ-BFA, G-BFA, and Ch-BFA by 32.857%, 8.538%, and 22.113%, respectively. Referring to Figure 10.8, the dominance rankings of the Pareto frontier and the individual solution rankings produced by the BFA variants are synchronous. The AER metric is utilized to measure the rate of exploration carried out by each of the approaches. The values of the obtained AER are shown in Figure 10.9.

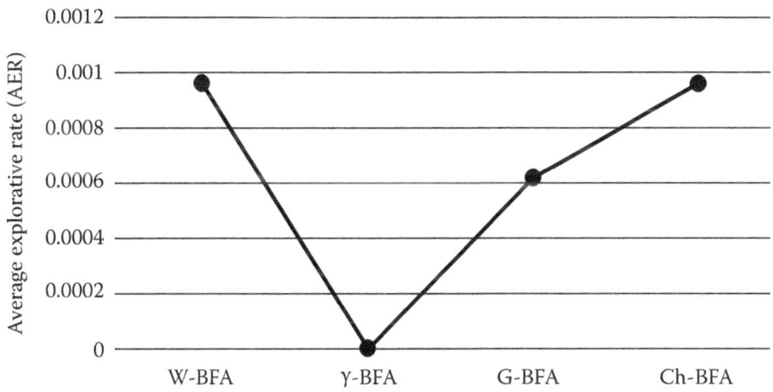

FIGURE 10.9 AER value for the BFA variants.

In Figure 10.9, the AER values are observed to be at maximum for the W-BFA approach followed by those of the Ch-BFA and the G-BFA. The lowest AER value was produced by the γ-BFA. It can be seen the AER value is related to the degree of dominance determined by the HVI. The W-BFA produces the most dominant Pareto frontier and the highest AER value. Similarly, the Pareto frontier generated by the γ-BFA has the lowest dominance and the lowest AER value while Ch-BFA and the G-BFA fall into the middle rank when measured with both the metrics.

10.8.2 PSO-Based Techniques

As with the BFA approach, each of the four stochastic engines was implemented with the PSO technique. The PSO variants equipped with the Gaussian, Weibull, and Gamma distributions are termed G-PSO, W-PSO, and γ-PSO respectively. The PSO variant which is coupled to the random chaotic generator is called the Ch-PSO. The entire Pareto frontier constructed using the G-PSO, W-PSO, γ-PSO, and Ch-PSO approaches are shown in Figures 10.10 through 10.13.

The overall dominance levels produced by HVI for each of the Pareto frontiers obtained using the PSO variants is shown in Figure 10.14.

Figure 10.14 shows that the W-PSO generates the most dominant Pareto frontier. The second most dominant frontier is obtained using the Ch-PSO followed by the G-PSO while the least dominant frontier was produced the γ-PSO. The frontier obtained using the W-PSO approach is more dominant than γ-PSO, G-PSO, and Ch-PSO by 60.253%, 54.321%, and 52.593%, respectively.

In Figure 10.11, the frontier produced by the W-PSO can be seen to have a good solution spread. The solutions in Figure 10.11 are well concentrated in highly optimal regions in the objective space. This critical feature makes the frontier obtained using the W-PSO more dominant as compared to all the other approaches employed in this work. The solutions in the frontiers produced by the G-PSO and the Ch-PSO (Figures 10.10 and 10.13) are seen to be concentrated in some regions and rather sparse in other areas of the objective space. For instance, Figure 10.10 shows that the solutions produced by the G-PSO are very sparsely

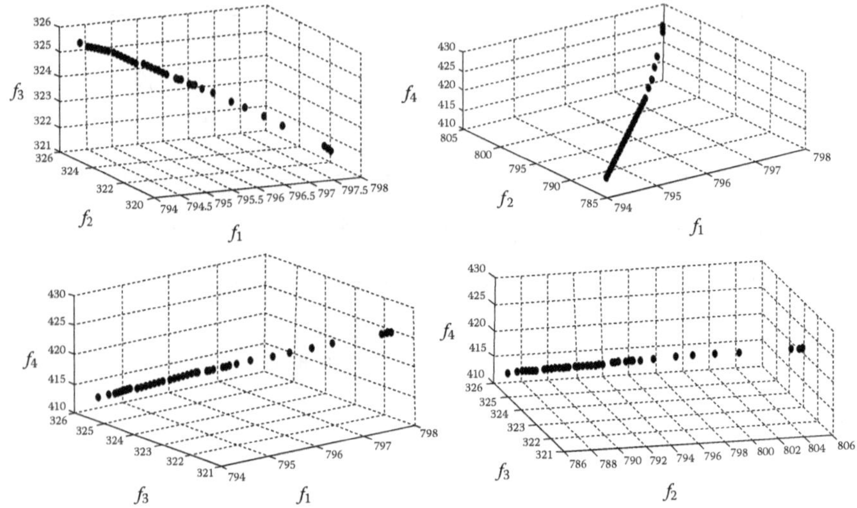

FIGURE 10.10 Pareto frontier obtained using G-PSO.

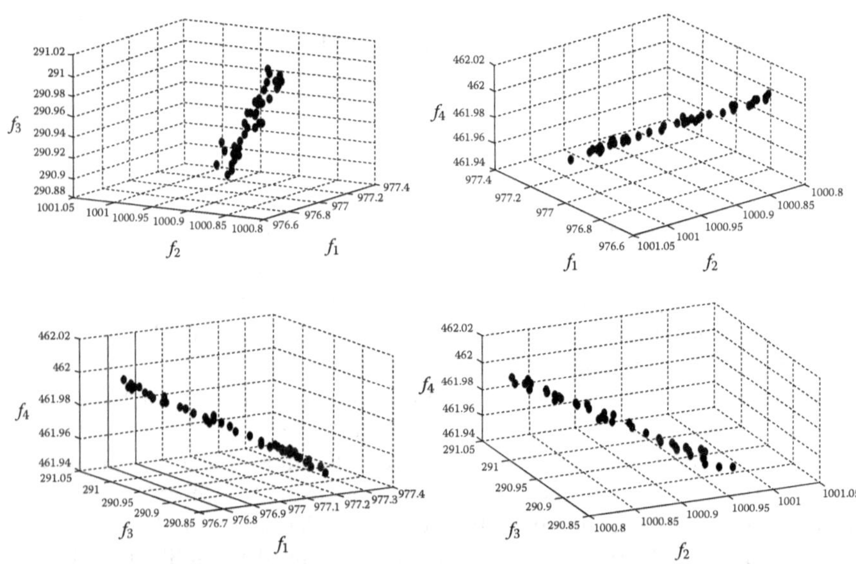

FIGURE 10.11 Pareto frontier obtained using W-PSO.

spread at regions with high values of f_1, f_2, and f_4 and low values of f_3. Regions with low values of f_1, f_2, and f_4 and high values of f_3 have high concentrations of solutions produced by the G-PSO technique. On the other hand, Figure 10.13 shows that the Ch-PSO technique produces highly sparse solution spreads at the extreme ends of the values of all the objective functions. The solutions on the frontier are mostly concentrated on the midrange of the objective functions.

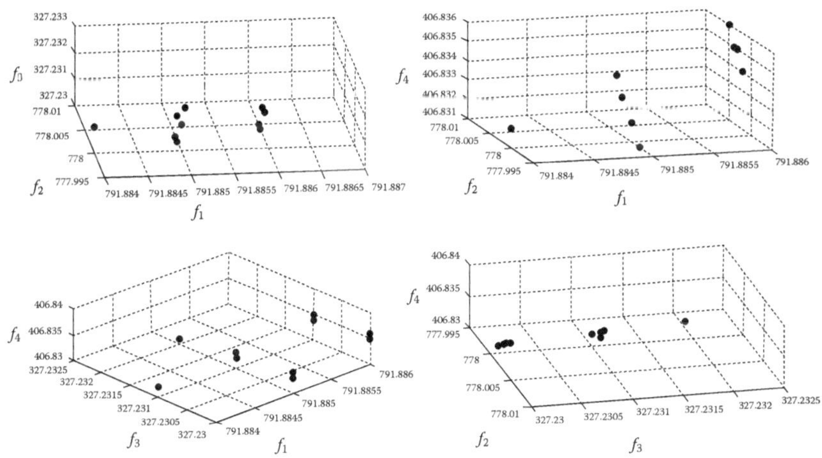

FIGURE 10.12 Pareto frontier obtained using γ-PSO.

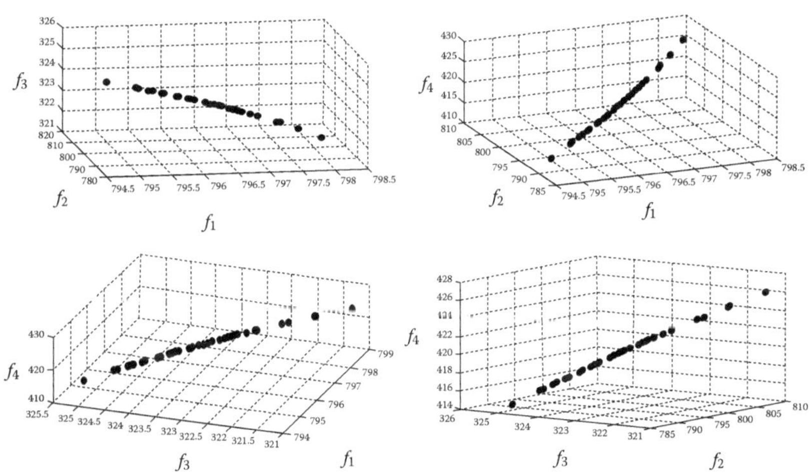

FIGURE 10.13 Pareto frontier obtained using Ch-PSO.

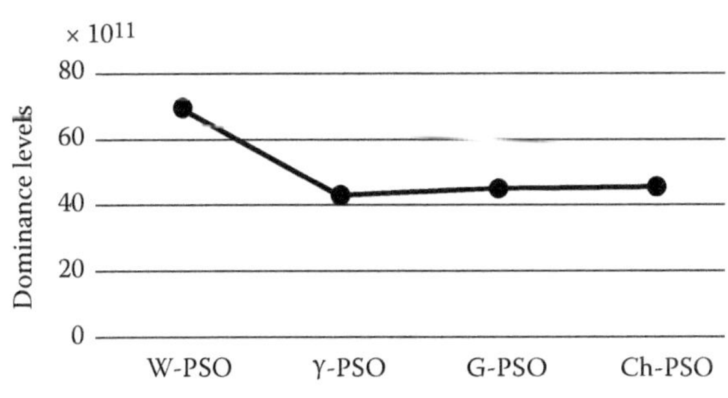

FIGURE 10.14 Dominance levels for the Pareto frontiers produced by the PSO variants.

Due to this localized nature of the solutions produced by the G-PSO and the Ch-PSO (Figures 10.10 and 10.13), it is more likely that the solution points miss the optimal regions in the objective space. Thus, as compared to the W-PSO, the Pareto frontier produced by both the G-PSO and the Ch-PSO have inferior dominance levels. The frontier constructed using the γ-PSO is extremely conglomerated (although having excellent coverage) to very narrow regions of the objective space (Figure 10.12). Thus, more of the optimal regions in the objective space are missed making the frontier by the γ-PSO the least dominant as compared to all the other techniques employed in this work. The values of the obtained AER are shown in Figure 10.15.

Figure 10.15 shows the AER values for the PSO variants to produce the entire Pareto frontier. It can be observed that the highest AER value was attained by the W-PSO approach followed by the Ch-PSO, G-PSO, and γ-PSO respectively. The AER value is seen to be related to the dominance levels identified by the HVI for all the techniques (Figure 10.14). Similarly, the Pareto frontier generated by the γ-PSO has the lowest dominance and the lowest AER value. The explorative rate is an indicator of how well the technique explores the objective space during the search operation. This effect influences the effectiveness of the approach in constructing a dominant Pareto frontier. The AER is very suitable to be utilized as an online metric since it does not require information from a completed search operation to be assessed. Therefore, the AER may improve the adaptability of the algorithm by iterative evaluation/correction procedures during execution. In this problem, the nadir point was chosen such that all the solution points produced by the PSO variants dominate this point. Thus, computational results evaluated using the HVI are independent of the nadir point selection.

To handle the MO feature of this problem, the weighted-sum scalarization framework is employed. Although the weighted-sum approach is very apt for Pareto frontier construction in MO scenarios, this approach fails to guarantee Pareto optimality (Shukla, 2007). The only setback with scalarization techniques such as the weighted-sum approach is that they may fail to completely describe the Pareto frontier if they are concavely structured.

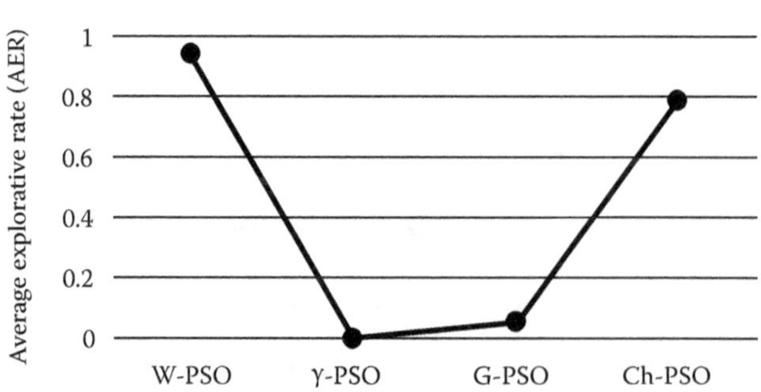

FIGURE 10.15 AER value for the PSO variants.

10.9 SUMMARY OF CRITICAL PERSPECTIVES

In terms of entire frontier dominance, the W-PSO narrowly outranks the W-BFA approach by 9.942%. This may be due to the minor algorithmic characteristics such as parameter settings or initialization point of the W-PSO which are more apt for the problem at hand as compared to the W-BFA. The introduction of the non-Gaussian approach in the conventional stochastic engine has shown very interesting results.

For both techniques, the algorithms equipped with a non-Gaussian stochastic engine outranks all the other approaches employed in terms of individual solution and degree of Pareto-frontier dominance. However, the γ-BFA and γ-PSO do not perform as well as the Weibull distribution-based algorithms or the conventional Gaussian-based approaches although they are equipped with a non-Gaussian stochastic engine as well. Therefore, although having a non-Gaussian stochastic engine may be advantageous, it is possible that the choice of non-Gaussian distribution employed in the solution method may be dependent on the type of solution landscape. Due to this dependence, the effectiveness of the computational approach with a particular type of non-Gaussian stochastic engine would vary based on the problem characteristics and type.

The AER metric provides good correlation with the measurements employed in this work. In future works, the AER metric could be employed in more computational approaches as an effective tool to direct the search effectively during program execution. In these numerical experiments, the AER metric was seen to correlate the measurements carried out with respect to the algorithmic exploration rate. The AER metric is an effective tool for directing the algorithm toward effective search operations at optimal regions in the objective space. Thus, the key points of this chapter are:

- Non-Gaussian stochastic engines significantly improve algorithmic performances.
- Information regarding the algorithm's explorative rate is crucial for understanding the dynamical interactions between the solution strategy and the problem landscape.
- The algorithm's explorative rate highly influences the level of dominance of the generated Pareto frontier.
- SI-based approaches are computationally efficient and suitable for complex real-world problems such as the resin-bonded sand core system.

REFERENCES

Afrabandpey, H., Ghaffari, M., Mirzaei, A., & Safayani, M. (2014). A novel bat algorithm based on chaos for optimization tasks. *2014 Iranian Conference on Intelligent Systems (ICIS)* (pp. 1–6), Piscataway, NJ: IEEE.

Ben George, E., & Karnan, M. (2012). MR brain image segmentation using bacteria foraging optimization algorithm. *International Journal of Engineering and Technology (IJET)*, *4*(5), 295–301.

Bensenouci, A. (2011). PID controllers design for a power plant using bacteria foraging algorithm. *Electronics, Communications and Photonics Conference (SIECPC)*, Saudi International, (pp. 1–6), Piscataway, NJ: IEEE.

Celik, A. N. (2004). A statistical analysis of wind power density based on the Weibull and Rayleigh models at the southern region of Turkey. *Renewable Energy, 29*(4), 593–604.

Corney, J. F., & Drummond, P. D. (2004). Gaussian quantum Monte Carlo methods for fermions and bosons. *Physical Review Letters, 93*(26), 1–4.

Cui, Z., & Gao, X. (2012). Theory and applications of swarm intelligence. *Neural Computing and Applications, 21*(2), 205–206.

De Pascoa, M. A. R., Ortega, E. M. M., & Cordeiro, G. M. (2011). The Kumaraswamy generalized gamma distribution with application in survival analysis. *Statistical Methodology, 8*, 411–433.

Elamvazuthi, I., Ganesan, T., & Vasant, P. (2011). A comparative study of HNN and hybrid HNN-PSO techniques in the optimization of distributed generation (DG) power systems. *2011 International Conference on Advanced Computer Science and Information System (ICACSIS)* (pp. 195–200), Piscataway, NJ: IEEE.

El-Wahed, W. F. A., Mousa, A. A., & Elsisy, M. A. (2008). Solving economic emissions load dispatch problem by using hybrid ACO-MSM approach. *The Online Journal on Power and Energy Engineering (OJPEE), 1*, 31–35.

Ireland, E., Chang, K., & Kroker, J. (2002). New horizon in no-bake binder technology. *AFS Transactions, 110*, 1–7.

Jakobson, M. (1981). Absolutely continuous invariant measures for one-parameter families of one-dimensional maps. *Communications on Mathematical Physics, 81*, 39–88.

Kennedy, J., & Eberhart, R. (1995). Particle swarm optimization, *IEEE Proceedings of the International Conference on Neural Networks* (pp. 1942–1948), Perth, Australia.

Kumar, D. (2013). Implementation of bacteria foraging optimization for color image quantization and its evaluation for various file formats. *International Journal of Computer Science and Communication Engineering, 2*(1), 36–42.

Lin, W., & Liu, P. X. (2006). Hammerstein model identification based on bacterial foraging. *Electronics Letters, 42*, 1332–1334.

Liu, Y., Passino, K. M., & Simaan, M. A. (2002). Biomimicry of social foraging bacteria for distributed optimization: Models, principles, and emergent behaviours. *Journal of Optimization Theory and Applications, 115*(3), 603–628.

Mamdani, E. H., & Assilian, S. (1975). An experiment in linguistic synthesis with a fuzzy logic controller. *International Journal of Man-Machine Studies, 7*, 1–13.

Mezura-Montes, E., & Herńandez-Ocãna, B. (2009). Modified bacterial foraging optimization for engineering design. In *Intelligent Engineering Systems through Artificial Neural Networks*. St. Louis, MO: ASME Press.

Mezura-Montes, E., Portilla-Flores, E. A., & Herńandez-Ocãna, B. (2014). Optimum synthesis of a four-bar mechanism using the modified bacterial foraging algorithm. *International Journal of Systems Science, 45*(5), 1080–1100.

Niu, B., Zhu, Y., He, X., & Zeng, X. (2006). Optimum design of PID controllers using only a germ of intelligence. *6th World Congress on Intelligent Control and Automation* (pp. 3584–3588), Dalian, China.

Nouri, H., & Hong, T. S. (2013). Development of bacteria foraging optimization algorithm for cell formation in cellular manufacturing system considering cell load variations. *Journal of Manufacturing Systems, 32*, 20–31.

O'Brien, W., Browman, H., & Evans, B. (1990). Search strategies of foraging animals. *American Scientist, 78*, 152–160.

Panda, S. K., Padhee, S., Sood, A. K., & Mahapatra, S. S. (2009). Optimization of fused deposition modelling (FDM) process parameters using bacterial foraging technique. *Intelligent Information Management, 1*, 89–97.

Parappagoudar, M. B., Pratihar, D. K., & Datta, G. L. (2007). Non-linear modeling using central composite design to predict green sand mold properties. *Proceedings IMechE B Journal of Engineering Manufacture, 221*, 881–894.

Passino, K. (2002). Biomimicry of bacterial foraging for distributed optimization and control. *IEEE Control Systems Magazine*, *22*(3), 52–67.

Rajinikanth, V., & Latha, K. (2012). Controller parameter optimization for nonlinear systems using enhanced bacteria foraging algorithm. *Applied Computational Intelligence and Soft Computing*, 1–12.

Ravi, K., Shilaja, C., Chitti Babu, B., & Kothari, D. P. (2014). Solving optimal power flow using modified bacterial foraging algorithm considering FACTS devices. *Journal of Power and Energy Engineering*, *2*, 639–646.

Roscoe, K. L., & Diermanse, F. (2011). Effect of surge uncertainty on probabilistically computed dune erosion. *Coastal Engineering*, *58*, 1023–1033.

Schneider, S. S., & McNally, L. C. (1992). Colony defense in the African honey bee in Africa. *Environmental Entomology*, *21*, 1362–1370.

Scott Turner, J. (2011). Termites as models of swarm cognition. *Swarm Intelligence*, *5*, 19–43.

Shirani, M., Härkegård, G., & Morinb, N. (2010). Fatigue life prediction of components made of spheroidal graphite cast iron. *Procedia Engineering*, *2*(1), 1125–1130.

Shukla, P. K. (2007). On the normal boundary intersection method for generation of efficient front. In *International Conference on Computational Science*, May, 2007 (pp. 310–317). Berlin, Heidelberg: Springer.

Sottinen, T., & Tudor, C. A. (2006). On the equivalence of multiparameter Gaussian processes. *Journal of Theoretical Probability*, *19*(2), 461–485.

Surekha, B., Hanumantha Rao, D., Mohan Rao, K. G., Vundavilli, P. R., & Parappagoudar, M. B. (2012). Modeling and analysis of resin bonded sand mold system using design of experiments and central composite design. *Journal of Manufacturing Science and Production*, *12*, 31–50.

Vijay, R. (2011). Intelligent bacterial foraging optimization technique to economic load dispatch problem. *International Journal of Soft Computing and Engineering (IJSCE)*, *1*(2), 2231–2307.

Weihua, L., Yingmin, L., Xueliang, Q., & Xiuling, L. (2008). Study on binder system of CO_2-cured phenol-formaldehyde resin used in foundry. *China Foundry*, *5*(2), 110–113.

Wernick, M. N., Yang, Y., Brankov, J. G., Yourganov, G., & Strother, S. C. (2010). Machine learning in medical imaging. *IEEE Signal Processing Magazine*, *27*(4), 25–38.

Wu, J., & Chen, E. (2009). A novel nonparametric regression ensemble for rainfall forecasting using particle swarm optimization technique coupled with artificial neural network. *Lecture Notes in Computer Science*, *5553*, 49–58.

Yildiz, A. R. (2013). Cuckoo search algorithm for the selection of optimal machining parameters in milling operations. *International Journal of Advanced Manufacturing Technology*, *64*(1–4), 55–61.

Final Remarks on Metaheuristics in Engineering

The recent rise in environmental concerns coupled with a rapidly evolving energy economy has marked the beginning of a new industrial phase. This new environment has rushed engineers and decision makers to drastically develop new engineering systems and significantly enhance existing ones. The direct effect of these efforts is the emergence of highly complex engineering systems. The main concerns that surface when dealing with such systems are: ensuring optimum performance, maintaining reliable and stable operations, reducing uncertainty during the design phase, and ensuring dynamical predictability. In the previous chapters, such issues occurring in real-world engineering systems were shown to be tackled with the aid of metaheuristic approaches. Applications from three critical industrial areas—energy systems, process, and material engineering—were presented and discussed. In those areas, the multiobjective nature of the problems makes it complex and difficult to solve. In addition to that, these problems are nonlinear, nonconvex, and multivariate. This book gives insights on the application of a solution framework when using metaheuristics to solve these problems. The framework consists of the effective implementation of the following notions:

- Efficient algorithm design
- Execution of multiple algorithms
- Performing algorithmic enhancements
- Evaluating solution quality via measurement metrics

Metaheuristic approaches are very suitable for handling problems which are highly difficult and cannot be solved using conventional or analytical methods. Metaheuristics have a great potential in engineering applications covering areas related to: modeling, machine learning, data analysis, pattern recognition, control, and optimization. Albeit effective, the approach itself may be indeterminate. In other words, the results from using metaheuristics may vary causing various inconsistencies. This is because the basic mechanism of metaheuristics is inherently stochastic. In other words, it begins by random initialization and then it progressively refines its solution until it reaches its desired results. If there are multiple desired results, then it may reach different results at every run. It should also be noted that unlike analytical or exact approaches, one does not know the metaheuristic algorithm's effectiveness prior to implementation. The same goes for the parameter setting of the metaheuristic which is usually tuned empirically during run time. The definition of the stopping criterion or fitness condition also plays a critical role. If the criterion is loosely defined, then the solution variance during each run is going to be high. Thus, due to the mentioned issues, a systematic solution framework is required for the implementation of metaheuristics.

In this book, strategies are introduced and applied to deal with the indeterminacy of the metaheuristic approaches, first being the idea of multiple program executions. Every time the metaheuristic is implemented, it is multiply executed resulting in a set of solutions. These solutions could then be ranked and the degree of variance during execution could be determined. Therefore, the engineer could account for the variations in the computational outcome.

Another approach is to attempt to understand the mechanisms involved in the metaheuristic by studying its previous implementation—application history. For instance, say, based on scientific literature that swarm-based algorithms have been successfully implemented in the past on economic power dispatch problems. Then if the engineer is dealing with a similar problem or a problem with similar characteristics, it would be wise to attempt solving it with a swarm-based technique.

On the other hand, understanding the intrinsic metaheuristic mechanisms could also be useful for: enhancing the metaheuristic algorithm and selecting the best metaheuristic strategy. Take for instance the engineering problem presented in Chapter 10. In that chapter various characteristics such as the algorithm's explorative rate and the dominance of the generated solutions were measured. By these measurements, the behavior of the BFA algorithms when endowed with various stochastic engines was ascertained. Therefore, a mechanistic understanding of the algorithm's behavior during implementation provides useful insights when enhancing it. Besides, its mechanisms identified from previous work could as well help when using a metaheuristic. For instance, it is known that certain multiobjective engineering problems with dual objectives are very suitable to be solved using non-dominated sorting genetic algorithm (NSGA-II) (Yusoff, Ngadiman, & Zain, 2011). Therefore, when faced with engineering problems with dual objectives which share certain properties as the problems previously mentioned, then the engineer would be better off equipped with some kind of metaheuristic equipped with a sorting mechanism.

The metaheuristic implementations in this book stress the notion of solution measurement. The sole reason for this is to obtain information regarding the algorithm's mechanism to understand its behavior and interactions with the engineering problem. Equipped with this understanding the indeterminacy of the metaheuristic implementation could be significantly reduced.

Parameter settings play a critical role in the effectiveness of metaheuristic approaches. Therefore, in this book, all metaheuristic approaches were tuned rigorously based on trial and error. Another popular way to tune the parameters is by using another algorithm (Eiben, Hinterding, & Michalewicz, 1999; Leung, Lam, Ling, & Tam, 2003; Xu, 2013). It should be noted that the parameter setting consists of all user-defined input parameters such as population size, dispersal coefficient, mutation operator value, and so on. Therefore, whichever method is employed (heuristic or algorithmic), rigorous and thorough parametric tuning is mandatory for the effective implementation of any metaheuristic.

As we walk into this new industrial phase, future engineering problems would be increasingly plagued by various complexities. When all else fails, metaheuristic approaches are the engineer's best tool at hand. In this book, the solution framework for effective implementation of metaheuristics in real-world engineering problems is presented. The authors hope that the procedures outlined in this book would be useful for engineers and practitioners dealing with similar problems.

REFERENCES

Eiben, A. E., Hinterding, R., & Michalewicz, Z. (1999). Parameter control in evolutionary algorithms. *Evolutionary Computation, IEEE Transactions*, *3*(2), 124–141.

Leung, F. H., Lam, H. K., Ling, S. H., & Tam, P. K. (2003). Tuning of the structure and parameters of a neural network using an improved genetic algorithm. *Neural Networks, IEEE Transactions*, *14*(1), 79–88.

Xu, G. (2013). An adaptive parameter tuning of particle swarm optimization algorithm. *Applied Mathematics and Computation*, *219*(9), 4560–4569.

Yusoff, Y., Ngadiman, M. S., & Zain, A. M. (2011). Overview of NSGA-II for optimizing machining process parameters. *Procedia Engineering*, *15*, 3978–3983.

Index

N

O

P

R

S

X

Y

Z

For Product Safety Concerns and Information please contact our EU representative GPSR@taylorandfrancis.com
Taylor & Francis Verlag GmbH, Kaufingerstraße 24, 80331 München, Germany

www.ingramcontent.com/pod-product-compliance
Lightning Source LLC
LaVergne TN
LVHW011127300926
849160LV00008B/548

* 9 7 8 1 4 9 8 7 1 5 4 8 5 *